CW01650825

A UNIFIED THEORY OF INFORMATION DESIGN

Visuals, Text & Ethics

Nicole Amare
University of South Alabama

and

Alan Manning
Brigham Young University

Baywood's Technical Communications Series
Series Editor: CHARLES H. SIDES

Routledge
Taylor & Francis Group
LONDON AND NEW YORK

First published 2013 by Baywood Publishing Company, Inc.

Published 2017 by Routledge
2 Park Square, Milton Park, Abingdon, Oxon OX14 4RN
711 Third Avenue, New York, NY 10017, USA

Routledge is an imprint of the Taylor & Francis Group, an informa business

Library of Congress Catalog Number: 2012012035
ISBN 13: 978-0-89503-778-7 (hbk)
ISBN 13: 978-0-89503-779-4 (pbk)

Library of Congress Cataloging-in-Publication Data

Amare, Nicole.
A unified theory of information design : visuals, text & ethics / Nicole Amare, University of South Alabama, and Alan Manning, Brigham Young University.
pages cm -- (Baywood's technical communications series)
Includes bibliographical references and index.
ISBN 978-0-89503-778-7 (cloth : alk. paper) -- ISBN 978-0-89503-779-4 (pbk. : alk. paper) -- ISBN 978-0-89503-780-0 (epub) -- ISBN 978-0-89503-781-7 (epdf) 1. Visual communication. 2. Written communication. 3. Peirce, Charles S. (Charles Sanders), 1839-1914. I. Manning, Alan (Alan D.) II. Title.
P93.5.A43 2012
302.2'2--dc23

2012012035

Table of Contents

INTRODUCTION 1

CHAPTER 1
DECORATIVES 27

CHAPTER 2
IMAGES 61

CHAPTER 3
DIAGRAMS 93

CHAPTER 4
INDICATIVES 115

CHAPTER 5
INFORMATIVE INDICES 141

CHAPTER 6
WORDS, SENTENCES, AND TEXT 165

CHAPTER 7
TOWARD A UNIVERSAL TERMINOLOGY AND
GRAMMAR OF VISUAL TYPES 187

REFERENCES 201

INDEX 207

Introduction

OVERVIEW

This is a book about visual forms of communication, including written text. These communicative visuals have a diverse range of forms and purposes, but we will show that it is possible to both describe and explain the major properties of diverse visual-communication forms and purposes within a common theoretical framework in much the same way that major properties of diverse chemical elements can be described with the same vocabulary of protons, neutrons, and electrons, all elements arranged in a single periodic table.

In other words, we will both describe and explain the major properties and purposes of written language, (i.e., words, sentences, and text) using the same vocabulary and framework that describes and explains the major properties and distinct purposes of decorative shapes and colors; images used both for decoration and object representation; visual signals that divide and call attention to other information on a page such as negative space, grid lines, and bullet points; visual signals that initiate action such as Web links and stop lights; diagrams that communicate relational propositions; tables organizing large arrays of information; and ritual-sequence visuals, such as amazon.com's shopping-cart metaphor, which initiate narratives or other ritually organized chains of action.

For those unaccustomed to thinking of written text as a visual form belonging to the same general class as the other visual forms listed above, consider, to begin with, what happens to a text's readability if we remove all white space and punctuation, visual signals of the same subtype as grid lines and bullet points, dividing and calling attention to adjacent information:

> wewillultimatelyidentifyevendeeperconnectionsbetweenfoundationalvisual
> designelementsandthegrammaroflanguageitselfbutobviouslythiswillfirstreq
> uireadiscussionofthosefoundations

We will ultimately identify even deeper connections between foundational visual design elements and the grammar of language itself, but obviously this will first require a discussion of those foundations.

No physicist or chemist today questions the value of a single theory that describes and explains a wide variety of phenomena, but oddly enough, the authors of this volume have frequently been asked why we are interested in advancing a unified theory of visual communication. The simplest answer is that we'd like to treat visual communication as a science, and seeking unified theories is just what science does. In more practical terms, a unified approach to visual communication enables us to teach visual design students relatively few things that will enable them to do relatively many things.

This is in contrast to the current situation in visual design pedagogy, where students have to essentially learn a distinct vocabulary and approach, and sometimes several vocabularies and several approaches for each distinct form of visual communication: one framework for decorative borders and backgrounds, another for decorative images, various competing theories of typeface choices, others governing color, still others for creating grammatical precision and clarity in text, and so forth.

Therefore, to illustrate the workings of our unified theory of visual information design, we will emphasize basic issues such as those that might be discussed with visual design students, how to make visual communication more effective and ethically sound, both in choosing one kind of graphic over another, and in fine-tuning the graphic form. Nevertheless, this is and can only be a preliminary discussion, outlining a unified theoretical approach to visual design, written mainly for academics who study communication and teach some form of information design, either textual or visual. These are our working definitions:

- An *effective*, *ethical* visual is one that serves attainable, sustainable purposes, purposes shared jointly by both the creator and the viewers of that visual.
- An *ineffective* visual lacks the form that is adequate for its chosen purpose.
- An *unethical* visual serves purposes of its creator that are not jointly shared by viewers, or serves purposes, even if jointly shared, that are not attainable or sustainable.

We will frame our discussion in terms of the work of C. S. Peirce.[1]

Peirce's philosophical and semiotic categories, we have found, provide the most effective template for constructing a unified theory of visual communication, a theory that serves not just to describe but explain phenomena as diverse as the emotional response to typeface forms and the grammaticality of sentences. We have found no framework other than Peirce's with so much descriptive and

[1] Peirce's surname is spelled e-before-i and pronounced like *purse*.

explanatory potential, and so we will use that framework *to more fully contextualize more-recent but less-unified thinking about visual communication.* The question is really not why we are using Peirce's framework so extensively for this current study, but rather why other academic studies of visual communication have neglected Peirce's work for so long. Some time ago, Iversen (1986) said it best when defending Peirce's semiotic system over others that are more frequently employed to explain visuals: "Peirce's richer typology of signs enables us to consider how different modes of signification work, while Saussure's model can only tell us how systems of arbitrary signs operate" (p. 85).

Born in 1839 and working privately on his philosophical system until the year before his death in 1914, Peirce had developed a significant reputation among scholars in the United States and Europe in the latter third of the 19th century (Ransdell, 1998). However, due to his ill health and personal idiosyncracies—both poorly understood and little tolerated by the academic culture of the era—Peirce was never able to hold a permanent teaching position and never received proper credit or consideration for much of his groundbreaking work (Brent, 1998).

With this book, we wish to add to the recent scholarship on Peirce that is just beginning to correct this historical oversight. We will draw primarily from Peirce's 10-category taxonomy of "signs" (i.e., meaningful forms) and his related theory of ethics.

Peirce is widely recognized as a founding thinker in several current disciplines (physics, mathematics, and artificial intelligence among them; see Burch, 2008). However, Peirce's work has not been fully explored in contemporary research on visual information and communication.

We undertake here as our primary purpose a more detailed application of the Peircean categories than has been previously attempted to problems of information design. Though Peirce's triads of sign-object-interpretant and icon-index-symbol have often been cited in passing, few have been aware that Peirce expanded these triads to describe 10 categories of sign (three kinds of icon, four kinds of index, and three kinds of symbol). Still less attention has been paid to the system of ethics that Peirce developed in conjunction with his theory of signs and their interpretation.

Peirce designated his 10 sign types using Roman numerals as shown below (Hartshorne & Weiss, 1935, paras 254–266).[2]

Ahead of his time as usual, Peirce's 10 sign types meant little to scholars of his day, but they effectively correspond with 10 distinct types of visuals common in modern discourse (exemplified in parentheses below). An understanding of each visual type leads to principles of effective visual design.

[2] Following standard practice among Peirce scholars, all citations in the posthumously published *Collected Papers, 1935–1958,* Hartshorne and Weiss, editors, are hereafter referenced in this format: (CP 2.254–264).

I. decorative icons (borders, typeface shapes, color, etc.)
II. image icons (photographs, realistic illustrations)
III. signaling indices (bullet points, gridlines, white space, etc.)
IV. action indices (stop signs, animation, Web links, etc.)

V. informative icons (diagrams, charts, graphs)
VI. reference indices (tables, sequenced menus, etc.)
VII. ritual indices (ritual signals, step-by-step procedures, etc.)

VIII. word-symbols
IX. sentence-symbols
X. whole-text-symbols

We will explain in later chapters Peirce's rationale for numbering the sign types as he did, along with their relative placement in his theoretical system. However, it should be especially noted that all elements of verbal language (Peircean types VIII, IX, and X) have important visual components in *written* discourse. All of these sign types, both the verbal and the visual, belong to a single theoretical system built on common principles. This insight from Peirce's work is particularly useful, though it too has not been fully explored previously. For example, Moriarty (2005) interprets visuals as signs based on Peirce's three types of sign relationships, but her discussion does not employ Peirce as a comprehensive and systematic means of classifying visual types.

We will examine the design properties of each of the 10 types as defined by Peirce's system. We will also focus on the useful communicative purposes of each type and how ethical issues arise when any visual type is misapplied in an unsuitable communication context, when purposes of the visual creator, the form of the visual, and the purposes of its viewers do not align. Newton (2005) states that "philosophers have so often linked the visual with the study of the ethical—sometimes positively, sometimes negatively" (p. 429), and we likewise explore in this book the human (mis)use of visuals.

Our discussion will be jointly grounded in Peirce's theory of communicative (i.e., semiotic) purpose and his epistemological ethics. In short, our goal is to improve understanding of visual communication by establishing the effective function of each visual type relative to specific communication situations (author and audience needs matched with media strengths and limitations of any given visual type).

Compromised Visuals: Illustrative Examples

The visuals in Figure 1, left column, illustrate the kinds of potential problems with visuals that our study will address. In public-domain examples gathered from various .gov Web sites, we find a range of information-design problems.

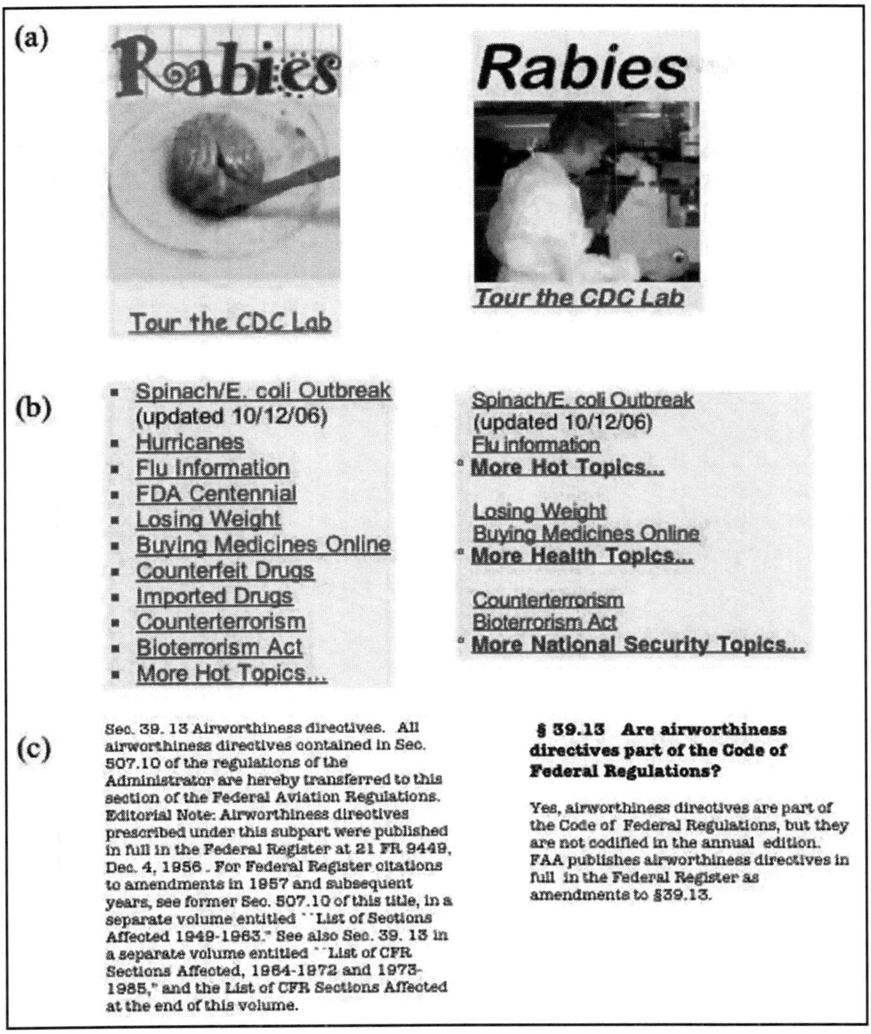

Figure 1. Decorative (a), indicative (b), and informative (c) visual problems and their revisions.

Visual design problems often (but not always) create ethical problems, in particular the undue effort imposed on users by careless design.

Figure 1, right column, illustrates possible revisions to deal with these problems. The Peircean model of formal design, purpose, and ethics will be used to both critique the information-design problems and motivate the revisions, all within a single theoretical framework.

For instance, the original visual from the Center for Disease Control (Figure 1a, left) deploys problematic typeface, color, and image choices (in addition to overly festive coloring not reproduced here). In brief, there is a jarring dissonance between the image of a pet's brain being dissected on a plate and the cheerful, party-like effect of the colored typeface header. Both typeface and image are out of step with the overall purpose of the site, intended to educate children about rabies, such that they can safeguard themselves and their pets against rabies infection. We may suppose that the designer of the original visual simply intended to create a fun, eye-catching display, but we will argue that the lack of care to keep visual elements consistent with the actual purpose of communication constitutes an ethical breach, even if the breach is unintentional. The illustrated revision of (a) might be further improved, as nearly any visual can be improved with an evolving understanding of viewers and purpose, but the revision shown at least avoids the most obvious problems in the original.

A comparable lack of care is indicated by this sampling of links on the Food and Drug Administration Homepage (Figure 1b, left). Here, links to various subtopics are listed out as bullets having no particular organization or order. Viewers are forced to visually examine each separate link. A more effective design would allow viewers to more efficiently consider smaller blocks of links organized by topic. Because the Web designer does not make the effort to visually divide and organize the links, each individual among the multitude of users is forced to expend considerable extra effort themselves. Again, this constitutes a kind of ethical breach, even if unintentional.

Likewise, there is an ethical problem inherent in carelessly formatted and therefore difficult-to-read text. In this case, the visually undifferentiated block of text and its visually reconfigured revision (Figure 1c) both come from the U.S. government's Plain Language movement. As with the other visual-display examples, information providers have an ethical responsibility to construct textual information in a way that serves the purposes of those who will view the information. Our point here is that the transformation of dense text into readable text is a kind of ethical responsibility and involves the same principles and techniques required to make visual information of all kinds more effective, more usable, and more ethical.

Of course, any well-trained visual designer could make the revisions shown in Figure 1 without applying a Peircean analysis. What is new in the Peircean analysis is first, a precise explanation for how it is that these kinds of improvements to visual communication are an ethical responsibility and second, that each of the revision strategies applied in Figure 1 can be explained within a single theoretical framework, such that designers trained in this new framework might do their work with more-focused and conscious awareness, with the potential for more-precise and reliable results.

Scope of this Book

This book is a preliminary discussion of a general framework for unifying the description of the major aspects of visual design with the major grammatical properties of language itself. We offer what is in essence a new typology and vocabulary for describing why certain visual forms are generally suitable for some purposes but not others. For this preliminary outline, we only require relatively simple examples of visual design of our own making or examples from the public domain (most of which are admittedly rather dated) in order to illustrate and define our terms.

Also, because this is a preliminary discussion, we cannot expect to prove conclusively, to every critics' satisfaction, that Peircean analysis of visual design purpose is as comprehensive and effective as we have found it to be. That determination can only be made, over time, by a large community of scholars who are willing to at least try the proposed methods of analysis for themselves, as we have illustrated them. As in any genuine academic inquiry, there will have to be an ongoing discussion about more-nuanced applications of our methods and any apparent counterexamples, as genuine challenges to the theory or only as cases where the main principles apply in less obvious ways, just as flying birds are not counterexamples to the general rule that things fall because of gravity.

There is no question that our specific interpretation and application of C. S. Peirce's thought can be further refined, improved, and extended by actual experimental investigation. Peirce himself was a scientist and would not have endorsed any other course. That investigation must begin, however, first with some iteration of an approach, which is all that we hope to describe in this book.

On the other hand, our discussion does not reduce to anything so simple as a list of prescriptions, such as "never use multicolored, high-variety typefaces." Rather, our approach simply determines, for example, that a high-variety, diversely colored typeface is (for most viewers) inconsistent and likely to clash with the average emotional response to an image of a dog's dissected and rabies-infected brain. Even in hypothetical cultures where dog brains might be served at festive occasions, a rabies-infected brain would surely be deemed inconsistent with the party-like emotions evoked by the typeface.

Such a determination is not any kind of prescriptive decree. Rather, we will be describing what we have found to be reliable generalizations about visual-design effectiveness and the typical alignments of form and purpose, where our investigation of these alignments has been guided by a particular theoretical perspective.

PEIRCE'S PRIMITIVE CATEGORIES AND VISUAL DESIGN

Loosely defined, we are undertaking here a study of the language of visuals. Scholars (for example, Kress & van Leeuwen, 1996) are working to develop complete visual-language systems, with vocabularies and interpretive processes at least as structurally detailed as the language of words, sentences, paragraphs, and texts. However, our approach is distinct from what is usually labeled as "semiotic" study. The more usual approach in such studies is to identify a specific image from art or advertising and discuss in some detail the possible covert messages and cultural implications embodied in the image. We recognize the need to understand how meaning might be made and revealed in imagery, but we will focus in this book on how to make graphics and graphical choices more effective for communication purposes that are explicit rather than implicit; that is, purposes that ought to be consciously shared by visual designers and visual viewers.

We will use as our framework the 10-class taxonomy of sign, the three-cornered typology on which it is based, and Peirce's related theory of ethics. This is not to say that all aspects of Peirce's thought are beyond criticism or improvement, nor are they to be accepted *a priori*. Rather, we are arguing here for relatively few specific aspects of Peirce's thinking, and our argument is based on the utility of specific Peircean concepts in describing and explaining various kinds of problems in visual information design, to more precisely characterize the difference between effective and ineffective design choices in several domains, as indicated in Figure 2.

Figure 2 illustrates the three key dimensions of Peirce's typology, (i.e., his general, three-cornered theory of classification). These typological dimensions generate several different kinds of visuals, depending on the strength of each dimension for each visual element. We would emphasize that these three dimensions are dimensions rather than absolute categories. Each visual category is defined by relative degrees of all three dimensions.

Peirce used this three-dimensional typology to generate the 10-class taxonomy of signs, both visual and verbal, but we will begin discussion with his three primary goals, the polar values of the three dimensions, as exemplified in Figure 3. Typeface choices, for example, are generally made to convey a particular feeling in a document, and so correspond with the first goal of evoking feelings, a goal that may also be served by borders, backgrounds, and images. Summarizing Robert Scholes's (1983) method for reading images, Helmers (2006) explains that our first reaction to visuals is an emotional one (p. 9). Visual strategies that serve mainly this first emotional goal we refer to as *decoratives*. Visual strategies that serve the second goal, mainly to provoke action, we refer to as *indicatives*. Bulleted lists, for example, specifically move an audience to the actions of separating, dividing, and contrasting otherwise

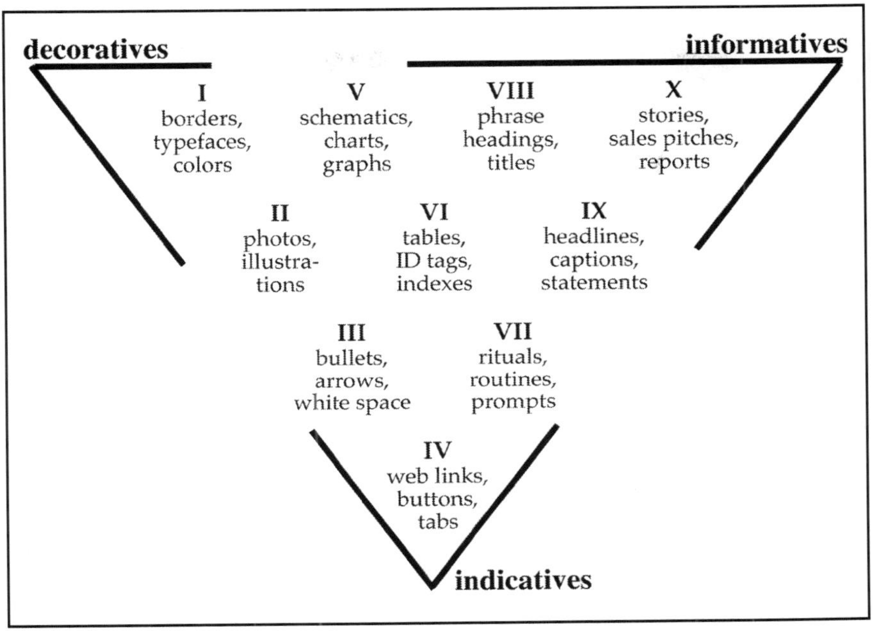

Figure 2. Visual-communication goals as defined by Peirce's three primary categories.

Peirce's various classes of sign are based on three fundamental dimensions, essentially three communication goals: to evoke feeling (decoratives), to provoke action (indicatives), or to promote understanding (informatives).

undivided statements in the flow of information. Note that the bullet marks themselves (*, -, >, etc.) do not inform in the technical sense of asserting propositions, but rather they indicate (i.e., point to) adjacent text statements that are informative.

Visual strategies that mainly serve the third goal, of promoting understanding by making visible those conceptual relationships more complex than mere pointing, concepts that can be paraphrased as true/false propositions, we refer to as *informatives.*

Words, sentences, and texts are, relatively speaking, the purest kinds of informatives, but visuals such as charts, graphs, diagrams, and tables have a concept-relational component, a yes/no-assertion component that can be distinguished from *whatever decorative and indicative properties they also have.*

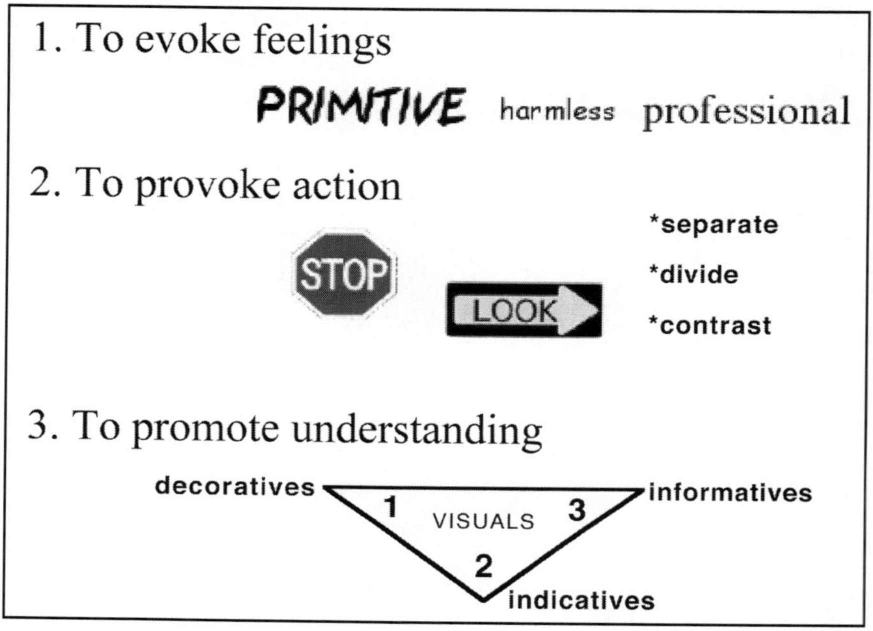

Figure 3. Primary visual-communication goals illustrated.

Typeface choices, as decoratives, evoke feelings. Bulleted lists, as indicatives, provoke action. Diagrams, as informatives, promote understanding.

Any visual design may of course incorporate a variety of elements. The Peircean typology will enable us to more precisely identify which specific features of an overall visual design serve which specific purpose, and to determine whether those features complement or conflict with each other. Under most circumstances, it is the informative purpose that governs the deployment of charts, diagrams, graphs, or tables, *even when they have subordinate decorative and indicative features.* That is to say that propositions asserted by these visuals are generally more pertinent than feelings or actions that they perhaps more immediately trigger. For this reason, various communication problems will arise if any undue emphasis is placed on the decorative or indicative aspects of primarily informative graphics. Some of these problems are practical, meaning that the visual will fail in its purpose. Some of these problems are actually ethical, meaning that the visual may serve a covert agenda of the author (for example, to demonstrate the visual designer's cleverness with decorative-design tools), an agenda that is not likely to be an agenda sought after by the audience. When visual designers show off at the expense of clarity and usability, this is not ordinarily thought of as an ethical problem, but the mismatch in author and audience goals is the very essence of an ethical breach according to the Peircean analysis.

PEIRCE'S EPISTEMOLOGICAL ETHICS

Peirce proposed a close integration of the visual and the textual. He postulated that "a universal art of rhetoric, which shall be the general secret of rendering signs effective, including under the term 'sign' *every picture, diagram, natural cry, pointing finger . . . numeral, word, sentence, chapter, book*" (from a 1904 MS published in 1978, pp. 148–149, emphasis added). We find here the assertion that both visual and textual effectiveness belong to a common system, built on and guided by similar principles. Our discussion of visuals here will consider the question of effectiveness as it pertains to ethics. Whether visual or textual, we can understand information effectiveness only in terms of communication goals: what the author means to accomplish with this or that visual effect and what effect an audience needs when presented with a text supported with visuals. Peirce defines rhetorical deployment of signs, visual or otherwise, in terms of effectiveness, and effectiveness must be defined in terms of needs, goals, or "ends." Consequently, there emerges a logically necessary connection between ethics and the effective achievement of ends (goals) in Peirce's system. "Ethics is the study of what ends of action we are deliberately prepared to adopt. That is, right action that is in conformity to ends that we are prepared deliberately to adopt. That is all there can be in the notion of righteousness, as it seems to me" (CP 5.130).

There is an easy misreading of Peirce here, made by many, to the effect that ends (goals) justify whatever means are used to achieve those goals. This is a misreading because Peirce here defines ethics as the study of effective actions relative to the deliberate (as opposed to a covert or unconscious) choice of goals. There is no ethics and no "righteous" action without deliberately chosen goals; there is no ethics, no standard for right action, unless precisely the same goals would be willingly chosen by a whole community, by all the parties to an action judged ethical or nonethical—the "we" Peirce uses in the passage above is significant. Newton (2005) explains this need for community consensus when determining what is good vs. bad: "ethics are determined by cultural and social mores that have become interpreted as right and wrong by groups of individuals" (p. 433). Thus, a message effective in immediate practical terms (i.e., utilitarian terms) might also be an ethical message but only so long as the goals or purposes that define a message's effectiveness are deliberately accepted by both the communicator and the audience.

This essentially deliberate and shared nature of ethical goals is an insight originating with Peirce, and it is missing in most contemporary discussions about ethics. This is also the major point that separates the Peircean ethic from a strictly utilitarian ethic of communication in particular and utilitarian views of right conduct in general. For example, marketing techniques encompass both issues of right visual communication and right conduct (in both a practical and a moral sense of right). Marketing techniques effective from a utilitarian standpoint can be highly suspect from an ethical standpoint because of the lack of shared goals: suppose that you see an advertisement (a specific type of indicative visual communication) for some gadget you've been wanting, very low-priced, at a local gadget store. You go to the store; the salesperson tells you they are currently out of the low-priced gadget (but does not tell you that the store never stocked more than one or two of these gadgets, which quickly sold). The salesperson offers to show you instead some comparable gadgets, higher priced to be sure, "but of much higher quality." This is the right-conduct aspect of the problem, but it relates directly to the ethics of the visual advertisement: salespeople may tell themselves that this strategy is ethical because the customer really is better off with a more useful, higher-quality product that they never would have bought were they not drawn in by the image of the cheaper product. This is the Utilitarian argument, but Peirce's analysis would designate this strategy as unethical, not merely based on the intuition of "wrongness" but based rather on an explicitly defined mismatch between the salesperson's initial bait-and-switch goals (to display visually an unavailable product but sell something more costly) and the customer's goals (to buy that low-priced product).

Through the Peircean analysis, we can explicitly connect bait-and-switch advertising of consumer goods to the ethical problem created by overly Photoshopped images of fashion models or other celebrities. This specific problem will be discussed in more detail in the chapter on images, but it suffices here to note that again the goals of the visual designer and the audience do not properly align: the visual designer intends to sell clothing or magazines by attaching their value to artificially enhanced images, images that the target audience will likely wish to emulate, but the Photoshopping visual designers cannot possibly share that emulation goal that they themselves deliberately created in their audience because they know that the images they have artificially created are impossible to attain in reality.

Moreover, it should be noted that ethical problems are not simply solved by full "transparency" between author and audience, either full awareness that author and audience goals differ or a full sharing by author and audience of goals, goals that can still be ethically defective because they are unattainable.

Peirce adds the further caution that deliberately and mutually chosen goals should also be ultimately attainable goals, sustainable in the long term by all members of a community. On these grounds, Peirce rejects hedonism (the goal of maximizing pleasure). It is not, in his view, an ultimately viable ethical system. "All motives that are directed toward pleasure or self-satisfaction, of however high a type, will be pronounced by every experienced person to be inevitably destined to miss the satisfaction at which they aim" (CP 8.140).

In other words, the hedonist agenda, even if shared by author and audience alike, always results in disappointment. It misses its deliberate goal. Drinking or taking drugs to forget one's problems always creates more problems, for example. Similarly, both visual designers and their audience might prefer visual information designs that are more attractive, more colorful, and more "flashy," even if these designs convey less information; but if more information is what is really needed, then the more attractive but informationally empty designs simply cannot be sustained long term.

We are thus able to classify as unethical any agenda that undercuts itself, even if the agenda is shared by a whole group: authors of an agenda and their willing audience alike. To take a contemporary example, the whole research division of a drug company might conspire to suppress evidence that a new drug has lethal side effects and to communicate among themselves with tables, graphs, charts, and other figures in support of a plan to minimize negative data. Again, this kind of practice reflects a faulty utilitarian ethic, subscribing to the idea that "ethics of the visual begins and ends with power, for power can determine whether something or someone is visible or invisible" (Newton, 2005, p. 434).

In the Peircean analysis, no single subgroup has sufficient power to make information invisible indefinitely if that information is relevant to the whole community's long-term purposes. Ultimately, if the drug comes to market and many people take it, the lethal truth will come to light. All the attempts at suppression, including the associated visuals, are seen in the long run as unethical exercises in self-deception and contrary to the goals of the larger community.

By this same argument, characterizing as defective any ethical system that cannot sustain or attain its goals, Peirce also rejects utilitarianism in its strictest form—that version of utilitarianism that requires all action or theoretical study to be immediately useful to self or society—because strict utilitarianism does not adequately consider the ultimate sustainability of its goals.

> Utilitarianism is one of the few theoretical motives which has unquestionably had an extremely beneficial influence. But the greatest happiness of the greatest number, as expounded by Bentham, resolves itself into merely superinducing the quality of pleasure upon men's immediate feelings. Now, if the pursuit of pleasure is not a satisfactory ultimate motive for me, why should I enslave myself to procuring it for others? (CP 8.141, citing Bentham, Jeremy A., A Fragment on Government, London: T. Payne, 1776, Preface, paragraph 2. This is a founding essay of the Utilitarian Movement.)

Referring again to the example of the Photoshopped imagery, the most aesthetically pleasing images may induce more immediate pleasure and happiness than realistic images would, for both visual designer and the audience. However, this pleasure cannot be sustained long term, as the general well-being of a society saturated with unattainable standards of beauty begins to suffer. Referring again to our marketing example, the salespeople may reason that the customer will be happier and most pleased with the higher-priced product. Even so, the sales method is doomed to ultimately fail in its ultimate goal: to sell lots of products and make lots of money. This is because, sooner or later, many customers will realize that they have been tricked into coming into the store, perhaps tricked into buying something they didn't want to start with, making the quality of the purchased product irrelevant. These people will not return to the gadget store, and they will advise all their friends to avoid shopping there. Ultimately, the store will go out of business. Its sales tactics, immediately useful as they were, failed in their ultimate goal.

Peirce's further complaint about utilitarianism is that the strict utilitarian inevitably discounts pure science, which is rarely useful immediately, but always has the potential to produce countless benefits in the end. Consequently, strict utilitarian methods, by short-sighted insistence on immediately applied theory and immediate social benefit, always fall short of the deliberate, long-term utilitarian goal.

> Now, to declare that the sole reason for scientific research is the good of society is to encourage those pseudo-scientists to claim, and the general public to admit, that they, who deal with the applications of knowledge, are the true men of science, and that the theoreticians are little better than idlers. (CP 8.142)

Peirce himself was an applied physicist, measuring slight differences in the Earth's gravity to cross-check altitude measurements for the U.S. Coast Survey, but he was also a pure theorist in mathematics and semiotics, and so he understood the immense but indeterminate value of pure research, which may pay dividends for centuries. The alternative that Peirce proposed to a hedonistic or strict utilitarian ethics might be termed an *epistemological ethics*. Ultimate truth is not known, but the disciplined thought and deliberate inquiry that leads to truth—the pursuit of truth itself—is a viable long-term goal that typically yields other goals as useful byproducts—both personal satisfaction and social utility—without the errors inherent in aiming at these lesser goals directly. "The only ethically sound motive is the most general one; and the motive that actually inspires the man of science, if not quite that, is very near to [the only ethically sound motive]" (CP 8.141).

This, in summary, is Peirce's epistemological ethics: in one way or another, all ethical goals require the open pursuit of truth and discourage the deliberate hiding of information that a community is deliberately seeking. Visual communication habits of a society may be seen as key indicators of the larger ethical commitments of that society. "The way we create visual messages, consume visual information, and store visual information affects the way we live, interact, and conceive of ourselves . . . Determining the best way to live is a core concern of ethics" (Newton, 2005, p. 440).

An epistemological ethics operates (or would operate) in the fashion-imagery example in this way: the actual appearance of models or celebrities is not deliberately hidden from the larger community. Clothing, exercise programs, diet advice, etc., are marketed with a range of accurate images that the larger community could, in principle, hope to emulate. It is up to individual members of the community to experiment for themselves and find those models that they can in fact emulate with success. An epistemological ethics operates in our marketing example in this way: the gadget store should carry adequate quantities of all the products that it advertises, especially the cheap ones. Salespeople are extraordinarily helpful when customers ask about the low-cost item, only pointing out that when the cheap gadget fails, or when customers themselves realize that they'd like a model with higher functionality, they are highly encouraged to return to the store for their upgrade. In this way, the customer is invited to reason and experiment for themselves to discover what level of gadget they really want. The salesperson is not simply telling the customers what they should want; customers themselves are part of the discovery process. When all parties participate in deliberate experimental inquiry—actions that serve the pursuit of truth—this is the very model of ethical action as Peirce understood it. "Now, thinking is a species of conduct which is largely subject to self-control. In all their features (which there is no room to describe here), logical self-control is a perfect mirror of ethical self-control" (CP 5.419).

As the passage above indicates, Peirce's characterization of thought is also key to understanding his ethics. He considered carefully disciplined thought, checked against the standards of logic and experimental method, to be a perfect model of ethical behavior in general.

Lacking the discipline of honest, self-critical science, people are more tempted to think carelessly, to accept unsustainable generalizations and rationalizations (that ultimately prove false). In contrast, with training in the discipline of truth-seeking, people are better prepared to overcome this temptation to careless thought and careless belief. Truth-seeking here can be interpreted in the Peircean analysis as the best general method for "relating means to desired ends" (Fulkerson, 1979, p. 343), and thus to "fail to have a consistent [truth-seeking] value theory or fail to let that philosophy shape pedagogy" is, as Fulkerson would have it, pure mindlessness (p. 347).

Likewise, lacking the discipline to think of their behaviors in terms of shared and sustainable goals (like shared and testable truth), people are more tempted to unethical actions, actions that ultimately always prove "false" by not being sustainable and "false" because such actions inevitably lead to ultimate results that no one would deliberately choose. Some end up enslaved by addiction rather than immersed in pleasure, and some end up in jail or bankrupt rather than wealthy and powerful, their actions not proving acceptable to the surrounding community. Likewise, visual designers end up with only a frustrated audience rather than an impressed audience, if the designers showcase their decorative design skills at the expense of information that their audience seeks.

With training in the discipline of ethics as truth-seeking, however, people might conversely be better prepared to resist temptations to behave less ethically, if disciplined to prefer "true" action (i.e., action that truly attains sustainable goals deliberately chosen by the whole community). True action, in this specialized sense, is ethical action by definition.

Ethical action cannot exist without community. A group of people have to choose common goals. "Peirce's epistemological theory [is] that all genuine knowledge is grounded in a community of inquiry that is jointly interacting with the real world" (Amare & Manning, 2009b, p. 288). Thought or, in other words, disciplined truth-seeking, must likewise distribute itself throughout a whole group. Thought, the key activity in the pursuit of truth, must be seen as an essentially community-based activity.

> Two things here are all-important to assure oneself of and to remember. The first is that a person is not absolutely an individual. His thoughts are what he is "saying to himself," that is, is saying to that other self that is just coming into life in the flow of time. When one reasons, it is that critical self that one is trying to persuade; and all thought whatsoever is a sign, and is mostly of the nature of language. The second thing to remember is that the man's circle of society (however widely or narrowly this phrase may be understood), is a sort of loosely compacted person, in some respects of higher rank than the person of an individual organism. (CP 5.421)

So on one hand, thought is a dialogue between different personas inside one mind. On the other hand, thought is also taking place, just as much or even more so, outside our individual heads in the ongoing conversation, in the exchange of texts between communicators and audiences in a community. We are back to the assertion that ethical communication must inevitably take "audience" into account, BUT not just in terms of being harmless, and not just in terms of accurate truth-telling, which are the focus of most current discussions of ethical communication. A Peircean epistemological ethic provides this additional dimension: that a communicator (like a gadget salesperson) cannot simply assert "truth" (e.g., "this is the product you really want") and expect that the audience can or even should automatically absorb what the communicator says as fact. Truth, by its nature, cannot be known immediately but rather requires long-term experimentation by a community of investigators before it can be accepted as such. In the short term, it is only the inquiry after truth that matters and that can be shared by communicator and audience. As noted above, visual information design can either facilitate this collective inquiry or damage it, depending on the competence and ethical orientation of the designer.

To summarize, ethical communication requires both the author and audience to be co-participants in a common thought process mediated by a set of shared and sustainable goals. Truth, to the extent that it is attainable, has to be validated or (re)discovered jointly by communicator and audience. This is the essence of Peirce's epistemological ethics, which we will use as our primary framework for discussing the ethics of visual information design. Ethical design centrally requires (a) deliberately chosen design goals, (b) shared with an audience, that are (c) attainable with the visual forms chosen. The discussion in this book will therefore focus on developing a framework for classifying design goals, matched with a framework for determining which visual design forms most often adequately achieve those goals. It happens that the classification framework outlined by Peirce best serves our purpose in unifying the discussion of design goals and design forms.

CHAPTER OVERVIEW

As noted above, Peirce's system is based on three basic principles, the logic of which we will develop in stages, through different chapters in this text.

Each of the outer "corner" types (I, IV, and X in Peirce's numbering above) can be described as the extreme manifestations of three basic communication purposes:

(Top left extreme) to decorate—to create a quality of feeling in the audience: borders, typeface shapes, color, etc., creating an overall feel for a document.

(Bottom center extreme) to indicate—to click a button, activate a program, turn a page, etc.

(Top right extreme) to inform—to promote in an audience further understanding of ideas, stories, sales pitches, reports, explanations, etc.

The other Peircean types (II–IX, discussed in chapters 2 through 6) all represent distinct intermediate forms, different combinations of the basic, primary communication purposes.

CHAPTER SUMMARIES

1. *Purely decorative (i.e., aesthetic) elements.* Shape, space, line, color, texture, and typeface choices are typically deployed with no clear purpose other than to evoke a feeling that the visual designer personally finds pleasing. Effective deployment of these visual elements requires that these forms create an appropriate emotional environment for viewers as well, and that these forms not interfere with the clarity of other elements, indicative and informative, in a visual.

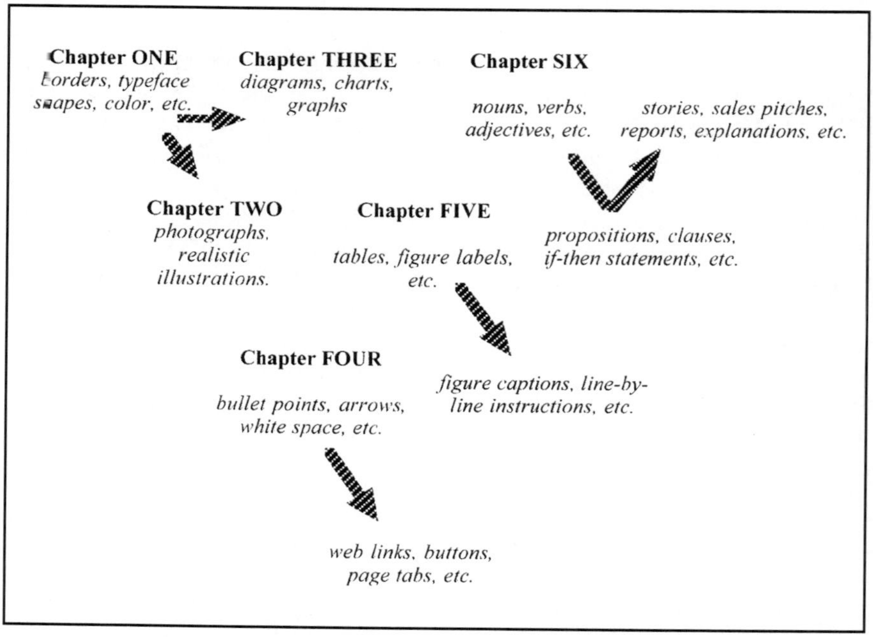

Figure 4. Progression of book chapters through the Peircean system.

Chapters 1, 4, and 6 will focus on what should be thought of as terminal points of a large triangle, the overall shape of Peirce's three-cornered system, a kind of periodic table of meaningful forms.

The Peircean categories offer a more specific vocabulary of purpose to evaluate whether standard visual design strategies actually serve a communication purpose shared by both author and audience. A mismatch in author and audience purposes may constitute an ethical breach.

2. *Decorative-indicative images.* Images may be largely decorative, but they also inherently evoke certain indicative properties, in that, in Peircean terms, they are one step down, in the direction of indicative purpose as shown in Figure 2 and Figure 4. Each image tends to provoke viewers to actively reference some specific object or kind of object in the domain of discourse. Images used for purely decorative intent (to merely evoke feeling) can therefore become problematic. Mismatch between an author's decorative and indicative intentions, or mismatch between an author's intent and the audience's expectations can produce

practical and ethical difficulties with images. Here again, the Peircean categories offer a more specific vocabulary of purpose to more precisely evaluate whether an image is being used effectively or whether an image choice is actually an obstacle to the larger purpose of a visual.

3. *Decorative-informative graphics.* Diagrams, charts, and graphs likewise have strong decorative (i.e., aesthetic) considerations, but they also should contain inherent informative properties. True informative properties, however, can be generated only under specific conditions: (a) redundant and extraneous details stripped away, (b) contrasts between critical details emphasized, and (c) generalizations represented in the graphic validated by collateral discussion. If these conditions aren't met, then the diagram/chart/graph will degrade into a mere image of information, a decorative-indicative substitute for information; that is, an impressive looking graphic that few or none in the audience can understand. This leads to practical failure of the visual and a possible ethical breach if the author has not exercised due diligence in making the graphic as informative as the audience requires it to be.

4. *Indicatives pointing to informative content.* Traditional visual design has not made an effective distinction between visual "contrasts" that are merely decorative (i.e., *varieties* of shape, space, line, color, texture, and typeface that create aesthetically "interesting" or "attractive" visuals) and those sharper visual contrasts that provoke physical action in viewers. In Peircean analysis, this is a critical distinction. In particular, strong lines, bullets, arrows, line-enclosed negative space, or radical changes in size, color, typeface, etc. all serve the primary function of physically directing audience focus. Any amount of decorative variety may be acceptable, but these strongly contrasting, strongly indicative elements can be effective only if used sparingly, in measured doses in any given figure or page. The Peircean analysis explains exactly why decorative variety can be fun and interesting, but indicative contrast is irritating and anxiety-producing if overused; to overuse visuals that are inherently indicative in effect is to shamelessly overwork the audience, both mentally and physically. Literal physical fatigue sets in, and the visual fails in its purpose. To drag an audience through a series of actions without genuine purpose has clear ethical implications.

5. *Indicatives containing informative content.* As noted in the previous chapter, there are severe limits on the number of raw, action-provoking visual elements that can be effectively placed in a visual design. But somehow, there is little difficulty when these same action-provoking visuals (lines, boxes, enclosed negative spaces, etc.) are embedded in ordered arrays of information such as tables and indexes. The Peircean analysis explains why this can be so. Tables and figure labels all will incorporate language to some degree, presenting numbers, quantities, terms, or other linguistic indicators in some kind of visual array (Peircean type VI). These arrays of indicators likewise provoke mental actions of focus, comparison, and contrast. However, informative content can moderate to

some degree an audience's visual tolerance for indicative detail. This means that the number of *ordered* rows and columns in any given table, or the number of labeled elements in a diagram, for instance, can be greater than the maximum effective number of *unordered* bullets, arrows, links, etc. on any given page or figure. Line-by-line instructions (First, fold tab A into slot B, etc.) belong to this same domain of informative indicatives (Peircean type VI). This is language that is always intended to translate directly into action. Although tolerance for detail can be greater with these types of visuals, there are still limits on acceptable detail, and likewise the author is still ethically obligated to move audience members to action only with their full understanding and consent.

6. *Informative language.* With all of previous chapters' discussions in place, we are in a position to demonstrate through Peircean analysis that language itself is organized under the same core principles as all visuals. In other words, the traditional distinction between visuals and texts is illusory and in many ways counterproductive. Effectively written text always incorporates other visual elements as well. Beyond the essentials of punctuation and spaces between words, Peirce's primary insight here is that language, either spoken or written, is not conceptually processed until and unless it is translated into visual terms. Information in the form of language is translated, at some cognitive level, into diagrammatic and tabular forms of information before that information can be understood in terms of perception (what we can expect to see, hear, feel, etc.) or in terms of action (how we should plan to act, based on information). Visual divisions of textual information—into subjects, predicates, sentences, paragraphs, and sections—all critically affect the translation of linguistic information into diagrammatic information and ultimately into perception and action. Whole stretches of text (Peircean type X) are invariably divisible into three parts: (a) a background of shared belief, (b) a question or issue that provokes doubt and drives the reading of the text forward, and (c) information that addresses the question. These natural divisions of information should likewise be marked out visually in paragraph breaks, sections breaks, and so on.

TOWARD A UNIVERSAL TERMINOLOGY AND GRAMMAR OF VISUAL TYPES

We will return, finally, to Peirce's overall system as outlined in the Introduction. We anticipate that we will not be able to explain to our audience some of the more advanced features of Peirce's system as it relates to modern types of visuals until after we have discussed numerous examples of each of the 10 types in some detail. With Peirce's logic of sign types then fully articulated, we can consider the implications of Peirce's model, along with the perspectives on language and mind embodied in it, for the field of visual communication in particular, but also for the general discipline of information design.

PEIRCE'S CATEGORIES AND THEIR GRAMMAR

Peirce's semiotic thought is vast in its details. Eight thick volumes of his writings were published by Harvard University between 1935 and 1958, and several other collections of his writings have been published since then, but many of his manuscripts remain unpublished still. As mentioned earlier, citations of Peirce's writing in his collected papers (CP) are to volume.paragraph (e.g., CP 6.201).

Despite its ultimate breadth and complexity, all of Peirce's thinking is generated from the core properties of three universal categories. Peirce sought to describe what are essentially the atoms or building blocks of all meaningful processes, atoms that would combine and create all the diversity of experience and thought in much the same way that three subatomic particles, protons, neutrons, and electrons, combine to make all the diversity of matter and materials that we encounter in the physical world.

Some might object that, because Peirce worked out his core categories of experience, meaning, and communication in the late 19th and early 20th centuries, his framework for describing and explaining visual communication could not possibly be relevant in the 21st century. It should be kept in mind, however, that the full periodic table of (natural) chemical elements was worked out by the end of the 19th century, and atomic physics worked out in the early 20th, and these frameworks are still very much in use. Peirce's fundamental categories are of the same order. The fact that scholars have overlooked them for most of the last 100 years has no bearing on their truth or usefulness.

Peirce hypothesized that everything that we perceive and think about is constituted by compounds of three fundamental categories. The essential qualities of these three categories are represented in Figure 5.

Firstness is defined this way: qualities of raw sensation or perception, like color, warmth, sweetness, etc, prior to their being identified as attributes of specific objects. These qualities are potentially, but not necessarily actually, realized in physical objects, as in a dream or hallucination. Beyond a biological nervous system, Firstness is exemplified by the potential energy of a boulder poised on a mountaintop that might at any moment roll down with a mighty crash but has not yet done so. In visual design terms, firstness is what defines decorative purpose: decorative forms like the color red consistently evoke some agitated kind of feeling, but otherwise the color, like all decorative forms, has a broad range of potential meanings, cultural associations, and interpretations that become fixed only when the decorative elements are "rolled downhill," as it were, into a specific indicative or informative context.

Secondness is defined this way: action and reaction, what is essentially the active response to some quality of firstness, like the shrinking away from pain, distaste, or the moving toward pleasure in reaction to a color or flavor. Beyond biology, when the boulder actually begins to roll and exhausts its potential energy

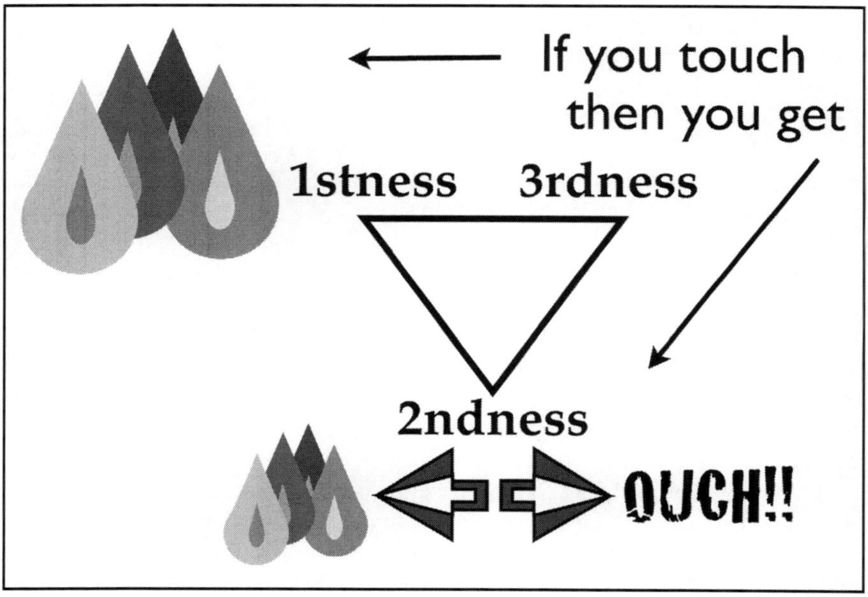

Figure 5. Peirce's primary categories and grammar.

Thirdness implies Firstness and Secondness because Thirdness relates a first and a second (here, fire and the reaction to it). Secondness implies Firstness because Secondness reacts to Firstness, but Firstness does not itself contain Secondness and Secondness does not contain Thirdness.

by crashing against other rocks in the valley, potential Firstness has been converted to actual, active Secondness. We are aware of a world of objects external to perceived sensations because of Secondness: we resist some of them in action, and some of them resist us in reaction—they are separate from us, separate from our personal expectations or desires, and this "otherness" is the essence of Secondness. In visual design terms, Secondness is what defines indicative purpose: indicative forms like Web page links provoke a physical action in viewers (click) and cause a definite action/reaction sequence of events, as do all indicative visual forms. Bullet points actively focus attention, indexes actively promote searching and locating, and so on.

Thirdness is defined this way: a pattern based on principle or habit, any regularity that we perceive when we interpret both sensation (Firstness) and reaction to the "other" (Secondness) and thereby create some representation of objects in terms of general categories, physical laws, rules of behavior, etc. Peirce generally construed Thirdness as the "meaning" of an indicative sentence (*X has properties Y*) and as a tendency to enforce a conditional (*IF X contains Y qualities, THEN perform action Z*). So for example, when one approaches fire, the perception of it is a mix of Firstness, qualities of heat and dancing color, but also Secondness, reactive sensations of pleasure or discomfort, actions in relation to the fire as object, depending on whether one feels cold or too warm already. Also, finally in the same experience there is Thirdness: one perceives that these particular sensations and reactions are of a general character—IF this X has the qualities of "fire," THEN EXPECT it to feel good on a cold day, to produce heat, to consume fuel and DON'T get too close or you can EXPECT to get burned!

These conditionals are thus a characterization of the meaning (Thirdness) of the concept of "fire," and indeed constitute the ultimate meaning of the word "fire," according to Peirce. Our sense of familiarity, regularity, habit, and of a rule-governed future we can anticipate depends on Thirdness, the interpretation and representation of percepts as rule-governed objects. In visual design terms, Thirdness is what defines informative purpose: informative visuals like diagrams must convey IF/THEN propositions (e.g., If this is tab A, THEN it must go in slot B) in addition to their decorative aesthetic qualities.

Peirce proposed that these same three categories are separately identifiable in any kind of experience, whether it be the simple perception of fire or something more complex like the interaction of desire, social codes, and choice.

To give one more illustration, if someone in a workplace setting, call her A, feels attracted to someone else in that workplace setting, call him (or her) B. This person A may have a desire to smile at that person B or reach out and touch B.

Love is the feeling, or Peircean Firstness. As long as that desire is not acted upon, it remains just a feeling of A, which potentially could go in many directions. If, however, person A actually caresses the arm of person B, this becomes a definite action or choice, which is Peircean Secondness, to which B must in one way or another react and participate in.

Secondness is inherently dual. There is no Secondness but only Firstness until A moves to actually touch and involve B.

Finally and third, there is a law of sexual harassment that is inherently triadic, external to A and B, but governing both. It informs A that she might be sanctioned for the uninvited touching of B, dismissed from the workplace perhaps, and so A either refrains or suffers the consequences of acting on her feelings. Knowingly or not, B is also affected by the law, either spared the touching in the first place or protected from its unwanted repetition. The information (Thirdness) does not erase the feeling or desire to touch, but it does determine the choice to touch.

Peirce's triad can be related to what Freud described as id, ego, and superego, but Peirce's categories have wider scope, being properties of everything external to human minds as well as everything internal.

The Grammar of Peircean Categories

The three Peircean categories have a hierarchical relationship that ultimately translates into a "grammar" (again, see Figure 5, with special attention to the caption). This grammar is essentially a set of restrictions governing how the categories relate to one another and how they can be combined. This grammar has two basic components: precedence and containment.

Precedence: Firstness (i.e., quality of feeling or potential) logically precedes Secondness (i.e., reaction to feeling or the actualization of potential). Both Firstness and Secondness logically precede Thirdness (i.e., the synthesis of Firstness, potential perception, the IF part of a conditional statement) and Secondness (actual results, the THEN part of a conditional statement).

Containment: Thirdness logically must contain both a Firstness element and a Secondness element; for example, an IF-THEN statement must have an IF part (first) and a THEN part (second), in addition to its full, third form. Secondness, conversely, does not contain Thirdness but must contain a Firstness; for example, a reaction must be a reaction to something. Firstness does not contain either Secondness or Thirdness, but only allows further elaborations of Firstness itself, that is, qualitative experience. For example, we may contemplate a field of wheat without reacting to it, but over time we will likely notice more and more qualitative details, color variations, movement in the wind, smells of dust and moisture, etc.

The Peircean grammar likewise manifests itself in the sexual harassment example, inasmuch as we would not have or need a sexual harassment law, for instance, without (first) strong feelings of desire and (second) active choices made from positions of power by actors driven by that emotion. We can, however, have Firstness (feeling or potential to act) without Secondness (action) or Thirdness (information or law). We can also have Secondness (action) without Thirdness (any law governing the action), but we cannot have Secondness (action) without Firstness (the potential or felt-desire to act).

In subsequent chapters, we intend to demonstrate the importance of this grammar to a proper understanding of effective visual design. For example, the grammar describes and explains the ill-formedness of those decorative forms that attempt to carry informative content (because Firstness does not contain Thirdness), but informative visuals (such as words and sentences) can function indicatively and decoratively (because Thirdness does contain Secondness and Firstness). The grammar is simple, but there are many important implications for visual information design.

CONCLUSION: AGAIN, WHY PEIRCE?

Peirce, more than any major communication theorist since his time, argued for the very point that we wish to make in this book. Text is a kind of visual, and all visuals are related in significant ways to the core structures of text. As was stated at the beginning of this introduction, the search for unified models and theories is the *sine qua non* of modern science. The point of Isaac Newton's theory of gravitation, for instance, was to unify the description of falling apples, cannonball trajectories, and planets in orbit around the sun. The point of the Periodic Table is to unify the description of all chemical elements, and so on. Peirce, more than any other thinker before or since, has provided us with a means to achieve a unified perspective on visual communication.

Those who suppose that there must be alternative unification models of visual communication are of course free to write their own unified theories of information design, but viable alternatives to Peirce's framework would have to prove equally comprehensive in its scope. In this present study, we can establish only a baseline for future work, to outline the Peircean approach and demonstrate the useful range of its application to visual design issues. Future work can expand on this same approach or challenge it, as researchers see fit.

For Peirce, visuals and text are only superficially different manifestations of the same *semiotic process*, meaning the exchange of feelings, actions, or information between minds by means of any kind of sign.

A *sign* is any visual form or any textual, tactile, or auditory form that conveys meaning.

> Logic, in its general sense, is, as I believe I have shown, only another name for semiotic ({sémeiötiké}), the quasi-necessary, or formal, doctrine of signs. I mean that we observe the characters of such signs as we know, and from such an observation, by a process which I will not object to naming Abstraction, we are led to statements as to what must be the characters of all signs used by a "scientific" intelligence, that is to say, by an intelligence capable of learning by experience. (CP 5.227, emphasis added)

In Peirce's system, the underlying meaning of any word, sentence, or longer text must be understood as a diagram or some other type icon, either visually or in some other sensory mode, either consciously or unconsciously.

> The only way of directly communicating an idea is by means of an icon; and every indirect method of communicating an idea must depend for its establishment upon the use of an icon. Hence, every assertion [i.e., verbal text] must contain an icon or set of icons, or else must contain signs whose meaning is only explicable by icons. The idea which the set of icons (or the equivalent of a set of icons) contained in an assertion signifies may be termed the predicate of the assertion. (CP 5.278)

In this book, our main purpose is to explain Peirce's system and demonstrate its specific relevance to contemporary communication in which visuals and text are so closely interwoven, to unify our understanding of both visuals and text in a framework of common principles. In 1986, Bernhardt helped us to "see the text"; similarly, we use Peirce to help information designers read text and images as part of a unified system of rhetoric.

Summary of Introduction

To summarize then, prior to our point-by-point discussion in subsequent chapters, the Peircean typology of visual design is defined by just these three basic categories: Firstness = feeling, Secondness = action, and Thirdness = information, which for our purposes can be presented as three distinct communication purposes:

1. (Peirce's Firstness) to decorate—to create a quality of feeling in the audience—borders, typeface shapes, color, etc., creating an overall feel for a document. We will call all such feeling-generating forms *decoratives*.
2. (Peirce's Secondness) to indicate—to provoke an audience to action, locating, dividing, classifying, etc.—Web links that can be clicked, action-activating buttons, page tabs that can be turned, etc. We will call all such action-provoking forms *indicatives*.
3. (Peirce's Thirdness) to inform—to promote in an audience further understanding of some idea—stories, sales pitches, reports, explanations, etc. We will call all such idea-promotion forms *informatives*.

We will demonstrate with numerous examples that Peirce was correct in his claim that the meaning of any textual information, if it is understood, has to be transformed through the mediation of diagrammatic forms into both perception (i.e., what we would see, hear, or feel if the information were true) and action (i.e., how we would act if the information were true). This insight is key to our analysis of visually configured texts.

CHAPTER 1

Decoratives

OVERVIEW

Visual designers, on a conscious level, most commonly use the basic elements of visual design, shape, space, line, color, texture, and typeface to evoke a feeling that the visual designer personally wishes to create. The designer must assume that viewers will find the design interesting, at least worth looking at, if not pleasing. However, the reality is that even artfully, even cleverly crafted visual designs can fail as communication unless the design creates an emotional environment for viewers that serves a larger purpose and is not just carefully crafted in terms of contrast, repetition, alignment, and proximity of visual elements. In short, however unified, organized, and interesting a visual looks, its design elements might, but should not, interfere with the clarity of larger purposes, indicative and informative, that a visual often needs to serve.

Our main focus in this chapter is to offer a more specific and comprehensive vocabulary of visual-design purpose in connection with the familiar tools of visual design form and the commonly understood effects of those forms, such as unity, organization, or interest. These so-called visual purposes (Williams, 2008, pp. 32, 50, 64, 80) are more accurately seen as "short-term" visual effects, not what designers and viewers jointly want to accomplish in the end, not purposes of the same order as the "long-term" purposes we have identified: decorative, indicative, and informative. This expanded vocabulary of purpose enables us to better evaluate whether standard visual design strategies are likely to serve a communication purpose shared by both the designer and viewers. As Williams correctly points out, once we know how to name certain visual features, we can more readily see them and start working with them (p. 11). In this and subsequent chapters, we will identify, through a systematic Peircean classification strategy, some demonstrable regularities of visual form effects in relation to long-term purpose. If designers (or design editors) become more aware of these regularities, they have some chance of consciously correcting designs that satisfy aesthetic considerations but fail to serve larger purposes.

We define decoratives as those visual elements that primarily evoke feeling, or what is otherwise known as "affect." Kress and van Leeuwen note how the affective qualities of visuals influence all messages:

> The production and communication of meaning is always affective and constitutive of subjectivities, in the domains we tend to regard as self-expressive (e.g., children's drawing and art) as much as in the domains we see as objectivating and impersonal (e.g., scientific and technical drawings). Even the maximally abstract modality of diagrams is an affective choice, by the very fact that it attempts to negate affect. (1996, p. 265)

Therefore, the decorative (i.e., affective) parameters of a visual should always be taken into account in visual information design. However, we find that even experienced visual designers can be tempted to overuse decorative visual elements at the expense of the overall communicative purpose. The decorative *chartjunk* so famously and often criticized by Edward Tufte (1997, 2001, 2003) is just one manifestation of this common problem.

The central problem here is that designers have to spend considerable time in creating a design, so much of the time, in fact, that their perception of the design (over hours and days) is highly likely to drift away from average viewer response, viewers who are only going to spend seconds, if not milliseconds, in evaluating the design and deciding if it meets their needs.

It is a problem that we might hope to remedy with a clearer set of tools to describe the distinct purposes that distinct visual strategies serve; tools to further determine how distinct visual elements can contribute to those purposes or interfere with them.

In a typical visual, decorative elements are mingled with indicative and informative elements, as in Figure 1.

The original version of Figure 1 (from a U.S. government Web page) was densely colored: deep red, orange, and pale peach against a teal background, but there are other significant decorative features as well: varieties of text size, shading, and placement, for instance.

The split-second response to Figure 1 tends to be something like, "*what a big fire!*" especially the colored version. We can track this interpretive misstep back to its source: the designer of this visual has attempted to indicate as well as decorate with the same visual elements of color, shading, size, and placement in space, as seen in Figure 1, where strong contrasts in these visual elements are used to indicate the contrasting kinds of storms. Meanwhile, the informative function of the visual is to assert the relative number of tropical storms, hurricanes, and major hurricanes in any given month during hurricane season.

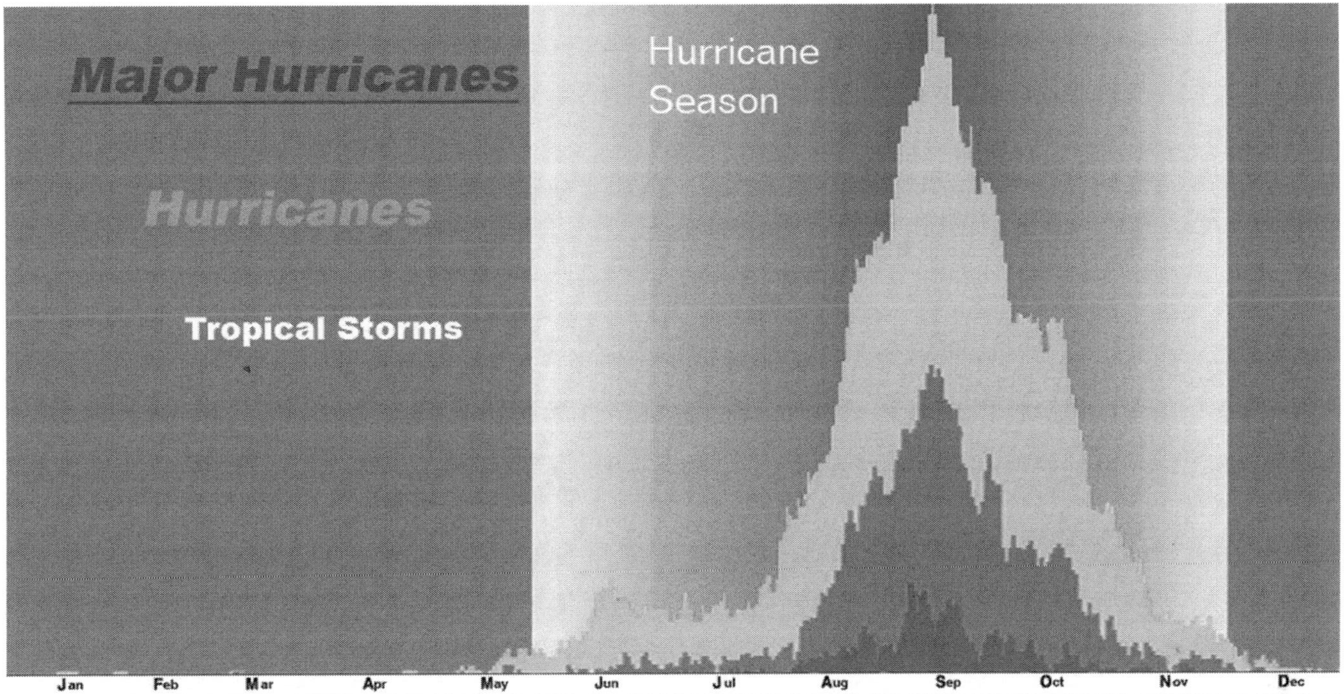

Figure 1. Significantly decorated graphic.

Decorative elements are typically mingled with indicative and informative elements. The challenge is to not let the decorative elements overwhelm key indicative points and information.

However, shading, size, and placement in this visual are *primarily* used to represent the *feeling* of relative severity.

The neutral shaded (blue-green in the original) background suggests the calm of months outside of hurricane season.

The light/medium/dark shading (yellow/orange/red shading in the original) suggests severity of risk during months of hurricane season, as well as the relative severity of storm types.

As is readily shown in Figure 2, however, the visual features described above as decorative are less effective and perhaps even misleading when they are simultaneously read as indicative, as pointing to key information in the visual.

Because the title "Major Hurricanes" is largest, highest, and darkest (= most red) in the visual, it naturally suggests the visual is primarily about major hurricanes. It takes a few moments of study to see, in the visual, that in fact tropical storms of minimal severity are much more common. The emotional impact of the visual forms tends to upstage the indicative and informative aspects of the graph. However, indication and information would normally be identified as the overall purpose of a line graph of this kind. Even so, the emotional message is the most prominent aspect of Figure 1, a message most literally translated as "HURRICANES! OH MY GOSH!" This message is most probably an unnecessary distraction in this particular context.

Decorative elements can be separated from nondecorative elements by a "removal test" (van Hooijdonk et al., 2007, p. 3). Removing decorative elements from a graphic will not change the propositions asserted (informative elements like *there are more major hurricanes in September than any other month*), and decorative removal will not change the objects, quantities, or ideas pointed to (indicative elements like *storm types*, *months*, and *quantities*), as in Figure 2.

When only decorative elements are removed or changed, then the *feeling* of the visual changes, but whatever is pointed to (indicated) in the visual, or asserted (as information) in the visual, does not change. Conversely, the indicative or informative properties of any color, size, or shading elements, i.e., those elements often but not always decorative are revealed by whatever has to be altered in a visual when color is removed, to retain the visual's indicative or informative properties.

For example, in Figure 2, all gradations in shading (or color) are removed, and consequently the label "Major Hurricanes" has to be connected to its graphed quantity by a line (instead of by the dark shade/red color). The labels "Hurricanes" and "Tropical Storms" must likewise be connected by lines.

Note, however, that the vertical order of the labels must change, and the contrasting typeface properties of the labels could change, with no loss of propositional information in the visual. Typeface form, point size, and even heading order prove to be the decorative elements in this case, directed at feeling rather than indication or information.

Readers can apply the removal test themselves, either mentally or in fact, to see that typeface choices are most typically decorative choices, meaning that

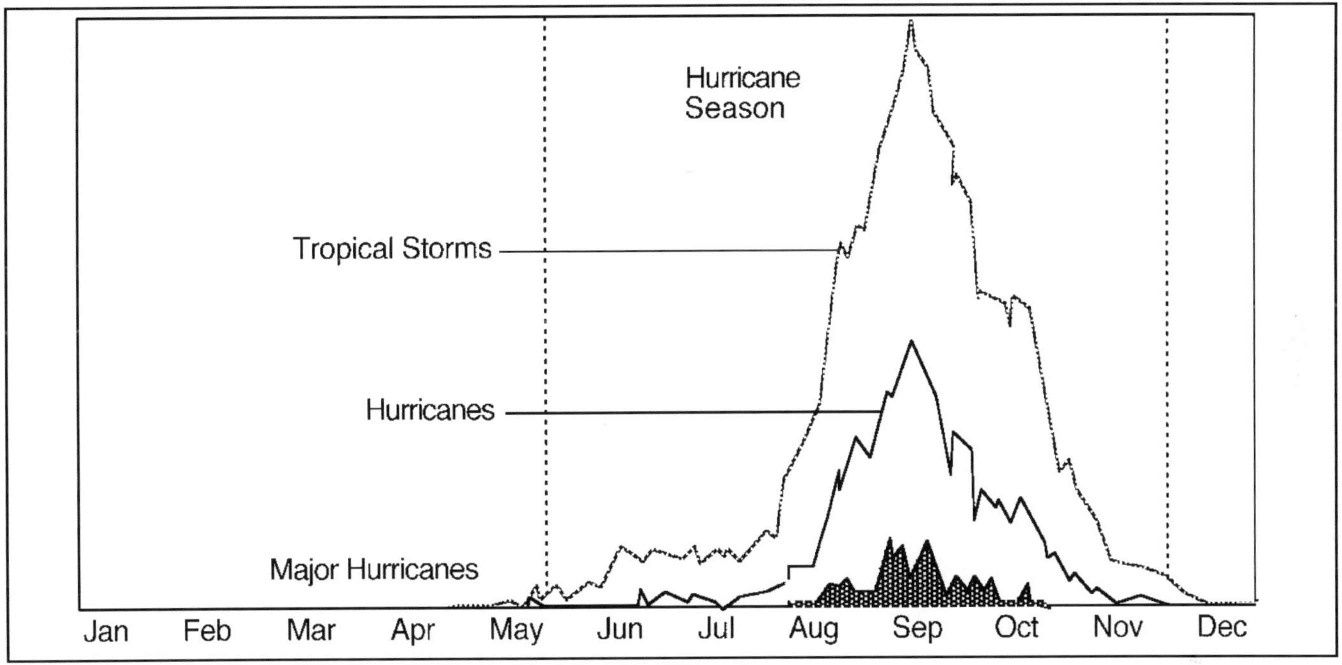

Figure 2. Relatively undecorated graphic.

Informative or indicative aspects of a visual design feature that is usually decorative (such as color or shading) are shown by those parts of a visual that have to be altered to retain information/indication when that visual feature is removed. Decorative aspects of a visual are revealed (typeface choices and the order and size of phrase labels) can also be revealed if change/removal does not alter what is indicated or asserted in the visual.

typeface choices determine the tone, mood, and feel of the text, including the felt level of readability (Brumberger 2003; Mackiewicz 2006; Mackiewicz & Moeller, 2004). These are all essential but nevertheless decorative features, as defined here in terms of their communication of feeling rather than information *per se*.

The main point here is not that visual design elements have a primary emotional effect, which of course is already well known. The point here is that emotional effects have been traditional visual design's almost exclusive focus. Visuals similar to Figure 1 are not hard to find, and they demonstrate a lack of awareness that a visual's emotional effects can and very often do interfere with other purposes in a visual. To avoid this outcome, fully competent visual designers ought to be able to separate, at will, the visual design elements that are driving emotional response from visual design elements that are indicating and informing. In many cases, the primary and legitimate purpose of the visual is to drive emotion, in which case the indicative and informative elements ought to be deliberately subordinated, but again, this is less likely to happen in any consistent way if designers aren't aware of the difference.

Advertisements tend to be highly decorated since "pathos" or feeling-creation is typically a key part of their persuasive task. On the other hand, technical information tends to be minimally decorated since object-identification and propositional information are the focus of technical communication.

However, even the absence of overt decoration will inevitably create a certain kind of "feel" in a visual: a sparse, stark, or plain feeling. As noted by Kress and Van Leeuwen (1996), "even the maximally abstract modality of diagrams is an affective choice, by the very fact that it attempts to negate affect." In our analysis, the term "negate" is perhaps too loaded a description. Rather, the plain style normally recommended for technical visuals is not so much negating feeling (affect) so much as the style is directed toward a deliberately *neutral* emotional field, a blank page in effect, upon which viewers are more free to choose their own emotional response to the information.

In advertisements, however, the main purpose of a design is to guide emotional response, and so the visual elements generally are and should be less neutral, more targeted to a specific emotional response.

Decoratives can enhance information, especially when used deliberately and ethically. The feelings evoked by visual forms can help or hinder emotional involvement, which "can have an important impact on how we interpret texts" (Gaskins, 1996, p. 386). People do not read with comprehension (or read at all) if they are emotionally upset (Ellis, Ottaway, Varner, Becker, & Moore, 1997). Thus, the emotional effects of decorative visuals can be positive or negative, depending on how effectively the decorative visual elements have been configured, to involve viewers emotionally or to distance them.

Well-used decoratives can help put readers at ease, or if the situation requires it, decoratives can make readers feel more agitated, motivated, and inclined to respond to a visual in an active way. A well-known example is the "Keep America Beautiful" (KAB) campaign against littering and pollution in the early 1970s. We can be sure that an undecorated version of KAB's perhaps most memorable ad (Figure 3) would have had little success. Conversely, their actual campaign made extensive use of decoratives (i.e., affect-creating, emotional elements) and was hugely influential in the early environmentalist movement (Figure 4).

Conversely, decoratives used carelessly (and thus unethically) can instead work to undermine information. A clash of poorly chosen decorative elements (Figure 5, a colorized, decorated version of Figure 3 from the Introduction) can create feelings of discord that inhibit comprehension.

In addition to the problem of emotional clash, information can suffer when decorative elements are overused to the point of distraction, or if they are used as lazy substitutes, or even as potentially deceptive substitutes (Figure 6) for the appropriate indicative and informative elements. Recall that the color and shape qualities of fire (*Firstness*) do not necessarily mean there is any physical fire danger (*Secondness*), and even less so do color and shape qualities of fire necessarily connect to a specific word or concept "*Fire*" (*Thirdness*). The range of distinct concepts that can be tied to color and form is indeterminately large.

By direct analogy, we cannot expect that decorative *Firstness* forms and colors by themselves (i.e., without external context, codes, or cues) would ever effectively communicate clear indicative messages (e.g., *GO HERE* = *Secondness*)

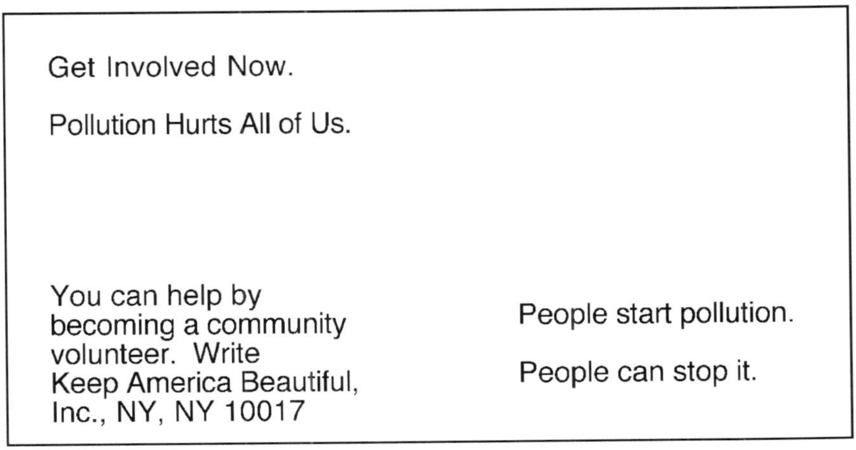

Figure 3. Undecorated persuasive graphic (less effective).

Informative elements alone are not likely to motivate in many cases.

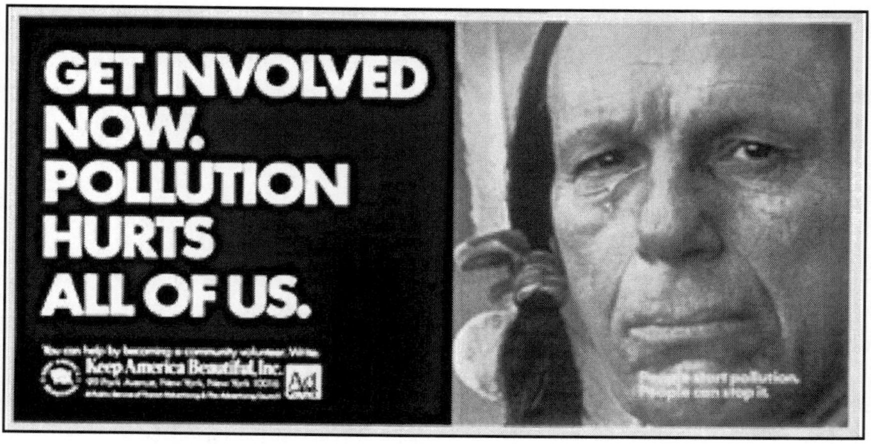

Figure 4. Decorated persuasive graphic (more graphic).

Keep America Beautiful's 1971 campaign was hugely successful, largely due to its decorative, feeling-evoking elements.

or clear informative messages (*X is greater than Y = Thirdness*). But someone can be misled by form and color into *feeling* like they see actual fire, and by this very same process, decorative forms in general can trick viewers into thinking they are getting information from the decoratives by themselves. In Figure 6, the decorative 3-D effects of the pie chart make some of wedges *feel* larger than they really are.

Pie charts are a common strategy for mixing decorative elements with information, information that often could be more clearly presented in simple tables. In the examples shown in Figure 6, depth and color effects can make smaller differences look, as for example, the difference between money spent on Capital Financing (23 million pounds = US$44M) and Highways/Transportation (32 million pounds = US$61M). Conversely, smaller quantities can be made to seem larger, as for example, the difference between employment growth in Education (4.7%) and growth in Personal and Other Services (6.1%) with the smaller wedge in a more prominent shade/color. (For discussion on distortion of 3-D pie charts, see Kimball & Hawkins, 2008, pp. 207–208.)

In this chapter, we will discuss commonly used visual elements that serve a primarily decorative function (without additional visual context): shapes, colors, textures, and typefaces. We will touch lightly on the decorative use of imagery, but images, with their partly decorative and partly indicative functions, will be discussed further in chapter 2.

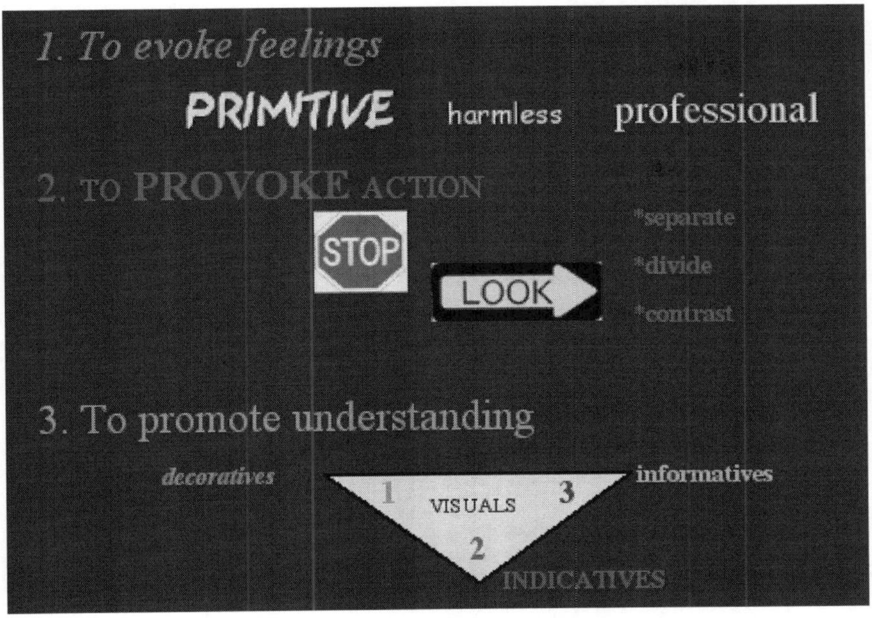

Figure 5. Ugly decorative, emotional-clash graphic.

Figure 3 from the Introduction, colorized version. A feeling of clash and unease is created by decorative elements, clashing colors, typefaces, or forms that create perceptual conflict rather than unity. Instead of reinforcing information, color here works as a distraction.

DECORATIVE DESIGN—PEIRCEAN PARAMETERS

According to Peircean analysis, each visual type has three key parameters:

- its basic form—whether the visual should be unified or have distinct, contrasting phases
- its referential process—how the visual acquires its meaning
- its interpretation type—what kind of meaning the visual form typically acquires.

Each of these parameters has, in turn, three different levels, corresponding with Peirce's basic categories of *Firstness*, *Secondness*, and *Thirdness*, as described in the Introduction. The properties that Peirce ascribed to *Firstness,* (i.e., qualities

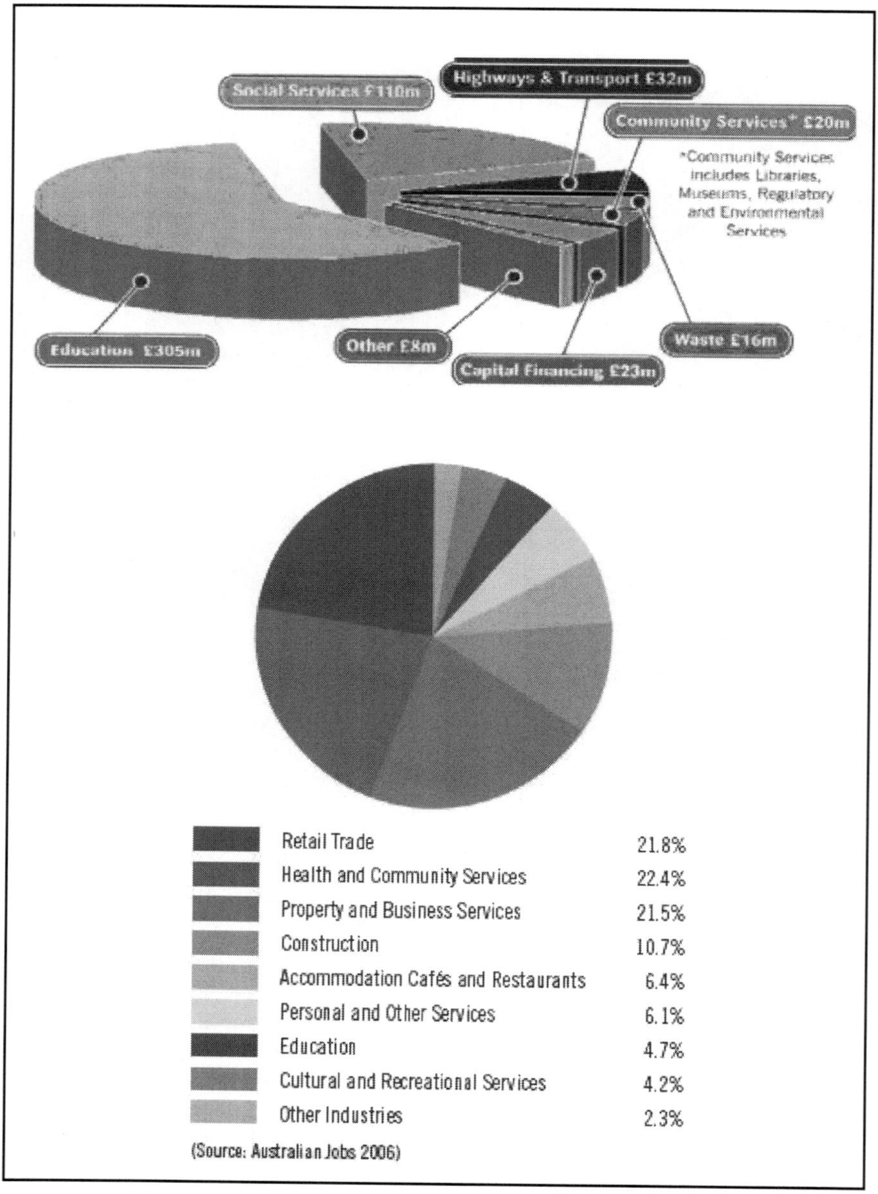

Figure 6. Potentially misleading decoratives, obscuring information.

Color and 3-D effects by themselves cannot indicate or inform, but they can distract.

of *unity*, qualities of *similarity*, and qualities of *feeling*) determine the behavior and useful application of decorative visual elements.

Basic Form

An effective decorative requires a form that creates qualities of *unity*: "This human preference for sensing unity where there is visual continuity appears to be analogous to that of sensing coherence where there is discourse relation" (Campbell, 1995, p. 17). All decorative elements in a single visual should work together to create a unified, holistic perception. In other words, decorative visuals must create a holistic gestalt. If a sense of gestalt, a unified whole, is not what the decorative creates, then a feeling of clash results (as in Figure 5), and the visual will probably fail in its decorative elements.

The same kind of clash effect is achieved in Figure 7a, where jarringly distinct colors and shapes are used, but the clash effect is minimized in Figure 7b. Note that Figure 7b uses essentially the same range of colors and shapes as Figure 7a, but they are configured differently in 7b, in accordance with gestalt principles all developed to describe the cognitive/emotional preference for unity, principles such as balance, symmetry, continuity, and closure (cf. Campbell 1995; Evans & Thomas, 2008; Riley & Parker, 1998; Sternberg, 2003).

In other words, the importance of formal unity in decorative visual design is well understood, and traditional visual design is already focused on teaching students how to accomplish it, but what is not well understood is how unified visual variety connects to larger purpose in visuals. It is understood, but only in a relatively vague way, that visual form should somehow meet audience needs, needs typically more complex than aesthetic harmony and visual interest. The Peircean analysis allows us to clarify how formal visual unity connects to the larger purpose of (iconic) reference by similarity and to construct an explicit framework for anticipating the range of emotional interpretation that a decorative form is (or is not) likely to evoke.

Reference Process

A decorative form gets referred to its interpretation through *qualities of similarity*. Decorative reference results from perceptual associations that viewers make between the form of the current decorative and any similar forms, previously experienced in emotional terms. The affective quality of a decorative, more often than not, depends to a large degree on viewers' prior cultural experiences. Decorative visuals can misfire if the author misjudges readers' cultural experiences and therefore the associations that readers will make.

Arguably, the advertising designers in both cases, Figures 8a and 8b, were attempting to associate the product with childlike fun and innocence, but they used very different means to make that association. Successful advertisers typically have to adapt their promotional work to different cultural markets, avoiding the pitfalls exemplified in Figure 8a, which relies on several associative

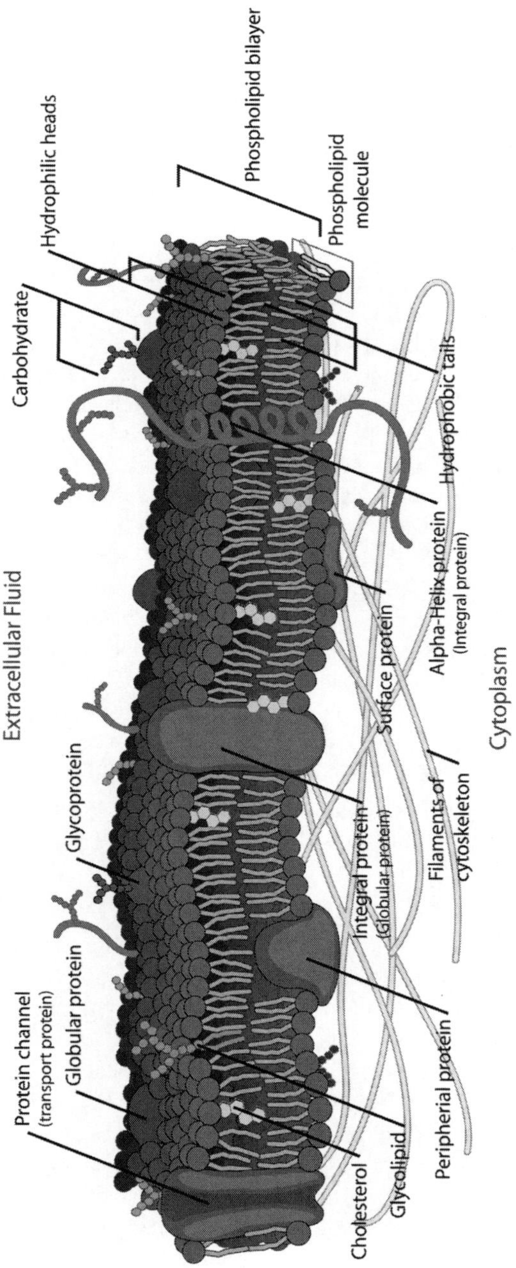

(a) http://commons.wikimedia.org/wiki/Image:Cell_membrane_detailed_diagram.svg.--PD

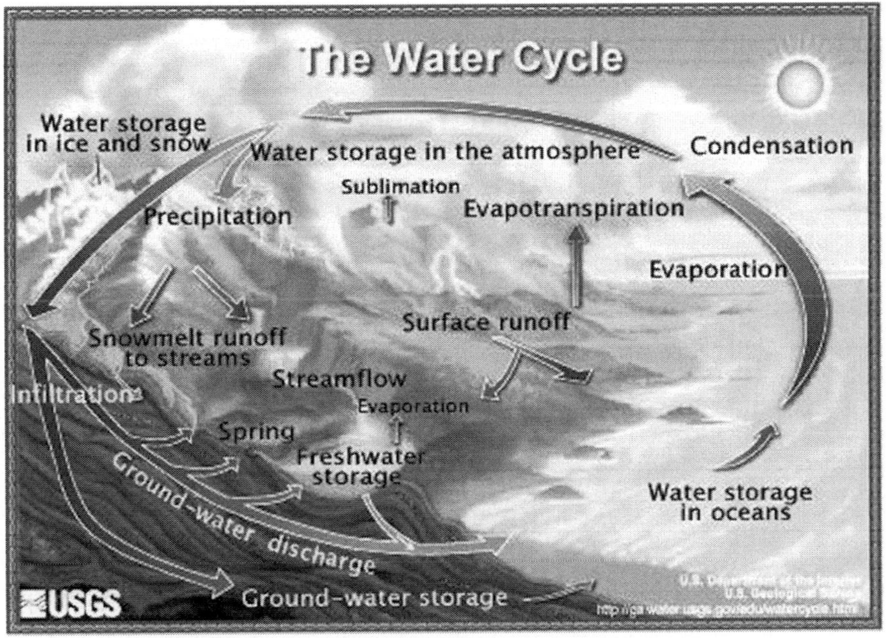

(b) http://ga.water.usgs.gov/edu/graphics/wcmaindiagram2.jpg

Figure 7a and 7b. Clashing decoratives vs. unified decoratives.

Figures 7a and 7b use essentially the same range of colors and shapes, but Figure 7a creates feelings of clash, while Figure 12b is configured in accordance with gestalt principles such as balance, symmetry, continuity, and closure.

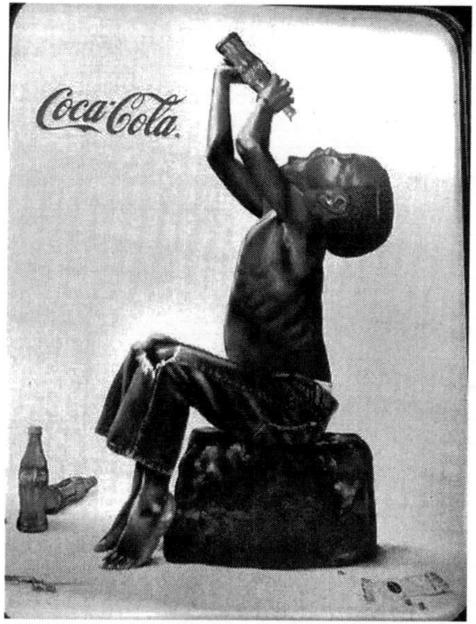

(a) http://www.neoc.ne.gov/edu/bhmreflect.htm
(Criticized for its racist connotations.)

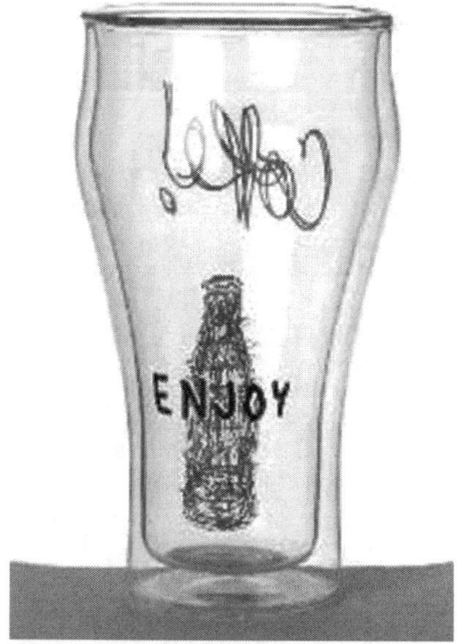

(b) http://www.cpse.gov/cpscpub/prerel/prhtml07/07268.html
(Discussed as a laceration hazard.)

Figures 8a and 8b. The same product advertised in different cultural contexts.

Advertisements rely fundamentally on associative connections that vary from generation to generation, from audience to audience, and from one culture to another.

connections that amount to racial stereotyping, which are of course unacceptable in current mainstream culture.

The marketing campaign illustrated in Figure 8b is only seen as problematic if viewers know that the commemorative glassware pictured here was found to pose a breakage hazard (the image taken from the government Consumer Products Safety Web site). With that awareness, the visual likewise takes on negative connotations or, in other words, negative associations based on perceived similarity, which is the fundamental reference process of decorative elements.

Interpretation Type

The core interpretations of decorative forms are *qualities of feeling*, meaning that, by definition, feeling or affect is what a decorative element can usefully communicate. We are pushing decoratives beyond their useful or "grammatical" range of application if we expect them, by themselves, to indicate or to inform with any degree of clarity. Pushing decorative forms into informative purposes almost invariably creates the wrong feeling as well. Referring again to Figure 7a, we see a visual that places too much of the information load on decorative elements, specifically color and formal variation, and the result is clash and confusion. Figure 9 eliminates those elements, shifts emphasis away from decoration, and limits itself to those gestalt features (e.g., symmetry and balance) that also work to structure information, structure appropriate for the apparent purpose of the graphic.

Summary of Peircean Design Parameters

The three Peircean design parameters for decoratives are summarized in Table 1.

In Peircean terms, all the decorative design parameters are First Level. This means that the three key aspects of decorative design each correspond with Peirce's first primitive category (Quality of Feeling; see Introduction). It can be useful, as a kind of shorthand, to refer to decoratives as being the "one-one-one" type of visual, written as 1-1-1. The first "1" refers to Form, the second "1" to Referential Process, and the third "1" to Interpretation Type, as in Table 1. This shorthand code locates decoratives relative to other visual types in Peirce's system of signs (Figure 10).

The Peircean system, as we represent it here, shows communication developing in several important ways from the top of the inverted pyramid to the bottom point and simultaneously from the left side of the pyramid to the right. Therefore, the location of decoratives in the 1-1-1 position (top and leftward in Figure 10) signifies at least three other critical attributes of this visual type.

- It is, from the standpoint of cognitive development, the first level of visual communication that novices will try to use and therefore misuse.

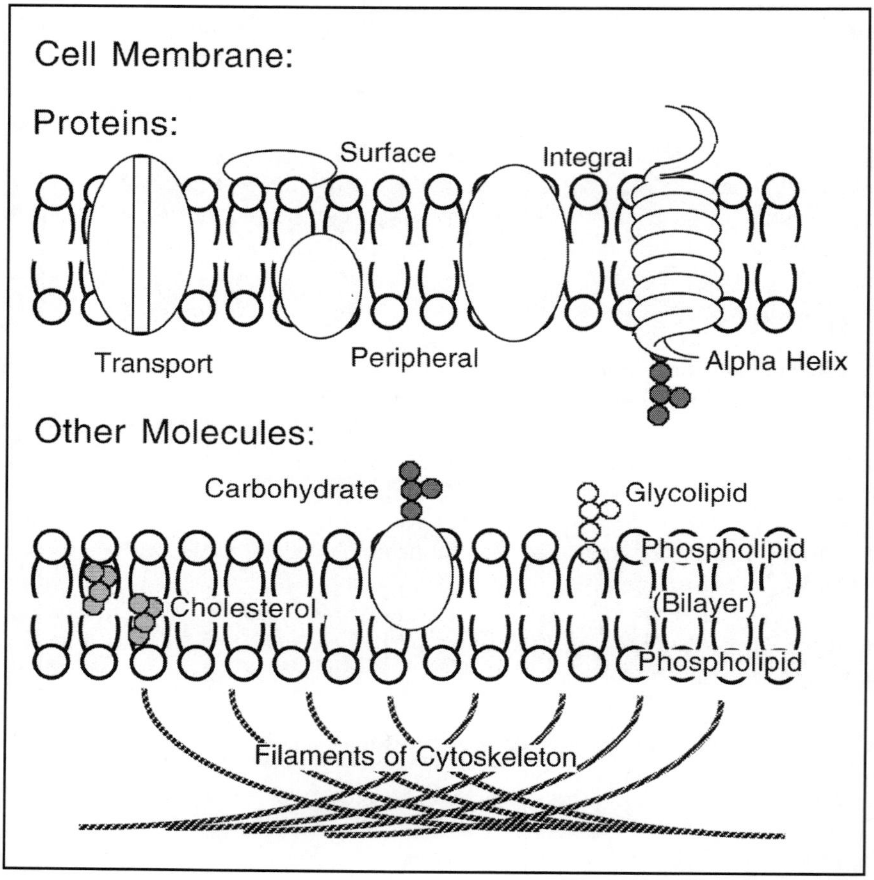

Figure 9. Informative reconstruction of an overly decorated visual (compare Figure 7a).

An overly decorated diagram (Figure 7a) is repaired by removing many, but not all, decorative elements.

- It is, from the standpoint of perception, the first level of visual communication that readers will most quickly notice in positive or negative terms.
- It is the first and therefore *least developed, least focused, least dense* medium of visual communication: potentially powerful in its immediacy but prone to multiple and vague interpretations in its message.

Table 1. Key Design Parameters of Decoratives;
Shorthand Code = 1-1-1[a]

Basic Form	Referential Process	Interpretation Type
Qualities of unity	Qualities of similarity	Qualities of feeling
Form Level 1	Reference Level 1	Interpretation Level 1
1	**1**	**1**

[a]In Peircean terms, all decorative design parameters are First Level: formal unity, referential similarity, and interpretive feeling. Hence, the shorthand code, 1-1-1.

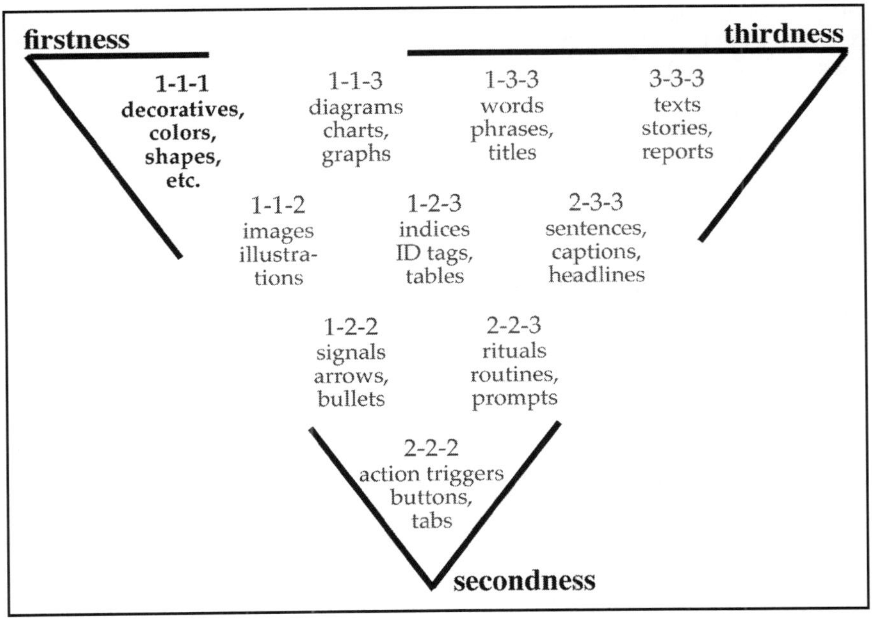

Figure 10. Decoratives (boldface) located relative to other visual types (muted).

Peirce's 10-class system of visual types repeated from the Introduction (Figure 2), with coordinate numbers added, 1-1-1, 1-1-2, 1-2-2, 1-1-3, etc., shorthand for the level of form, referential process, and interpretation of each visual type.

DECORATIVE PURPOSE— APPLICATIONS AND LIMITATIONS

There are many legitimate uses for a visual mode that is immediately accessible, emotionally evocative, and yet informationally vague, but we must also keep the limitations of this visual mode clearly in mind. If the primary purpose of a visual is to share a feeling, that is, the celebratory feeling of a party invitation, the somber feeling of a sympathy card, or the excited feeling of an eye-catching advertisement, then the very prominent decorative elements are appropriate. However, if the primary purpose of a visual is to indicate or inform, then only muted, subtle decorative elements may be appropriate, and then only if these decorative elements are kept below the threshold of distraction.

Shape and Texture

Decorative forms, such as the shapes and textures that make up borders and backgrounds, depend on a general sense of *unity*; their meaning develops through a process of association by *similarity*, and their interpreted meanings constitute somewhat vague *feelings* that can nevertheless be classified, for example, pleased vs. uneasy feelings, agitated vs. peaceful feelings, or resolved vs. aroused feelings.

A minimum line or the background (i.e., the negative or "white" space) in a visual can serve a nondecorative (indicative) function, serving to separate or differentiate one visual element from another. However, as soon as extra features like weight, color, or texture are added to that indicative line or background space, then decorative parameters of formal unity, reference-by-similarity, and interpretative feeling have to be weighed in the choice of those extra features.

Recall again Figure 2, p. 31, a relatively undecorated graphic, in contrast to the decorated version, Figure 1, p. 29. The more decorated version entails considerations that the undecorated version does not; for example, whether colors in the red-orange-yellow range transition too abruptly from the cyan background (threatening the unity of form), and whether red-orange-yellow will evoke the wrong associations for viewers (threatening reference by similarity), distracting them with irrelevant associations to bonfires, etc., as opposed to alerting them to the dangers of hurricane season.

Unity: Decorative shapes and textures create an expectation of unified form, meaning that some motif established by one area of the visual must continue across the whole. If that continuity is violated, if the definitive motif is interrupted, then a feeling of clash is likely. The border or background then becomes problematic, as shown in Figure 11.

In other words, gestalt principles apply particularly to decorative elements, and other visuals with level-1 form, as discussed, for example, by Paul Lester, "the law of continuation rests on the principle, again assumed by Gestalt psychologists, that the brain does not prefer sudden or unusual changes in the movement of a line.

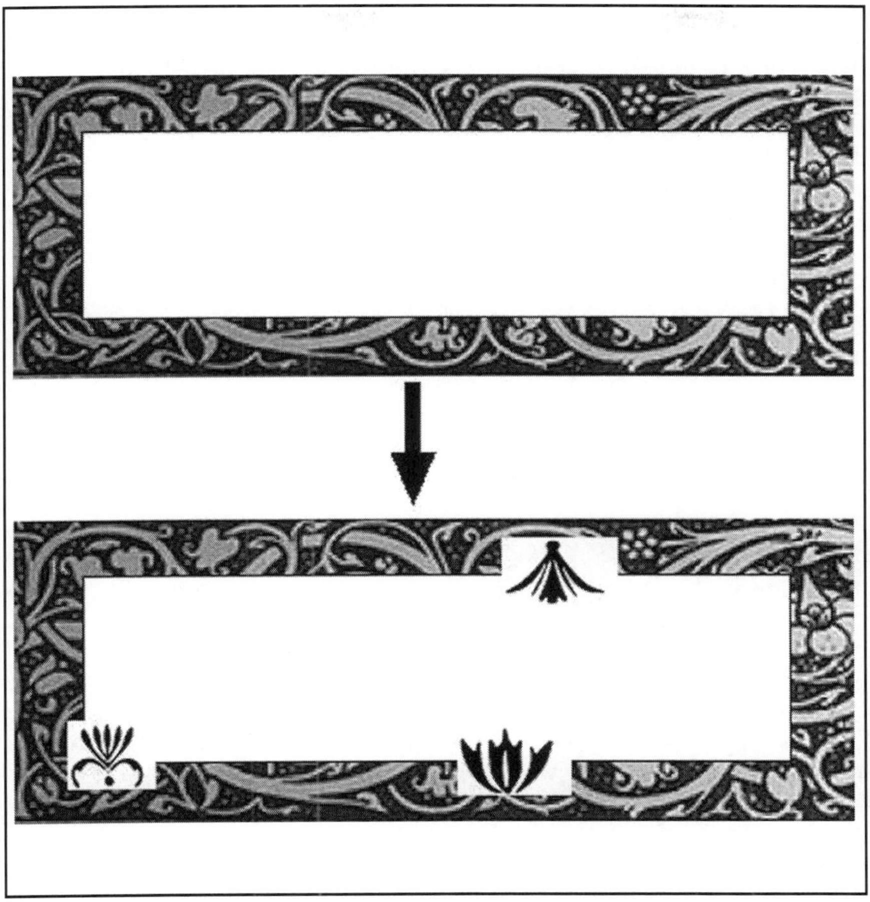

Figure 11. Interrupted border or background patterns are problematic.

*Gestalt principles are essentially unity principles (Level-1 form); if unity is violated, a decorative motif is (in Peircean terms) grammatically ill-formed: its Peircean code designation would be *2-1-1, meaning a visual that refers by similarity and is interpreted as feelings, but that has disruptive contrast that breaks overall unity.*

In other words, the brain seeks as much as possible a smooth continuation of a line" (2003, p. 51). However, we would further emphasize a point that will be clearer later in this book, which is that while gestalt principles do hold most strongly for decoratives and to some degree for any visuals with level-1 form (i.e., images, diagrams, and signals), the gestalt requirements of unity, continuity, etc. do not hold for all the visual types.

> More than seeing a whole as more than the sum of its parts, Gestalt theory describes why we perceive things the way we do and why what we see as a visual does not necessarily mirror reality (in fact, it usually doesn't). Gestalt theory alerts us that we do not always see things in their natural state. (Amare, Nowlin, & Weber, 2011)

Visuals often must specifically affect reality. For example, the title of a book needs to physically grab viewer attention. Hence, the visual design must include elements that are not merely about evoking feeling but provoking action, and it is common here to use two highly contrasting typefaces in one visual design, the effective purpose being to create a focal point of viewer attention, which is an indicative, reality-affecting function. Indicative and informative visuals often specifically require abrupt breaks in form, as will be seen in later chapters. Assuming, however, that decorative emotion is the primary goal of a visual element, in essence a step back from pointy, nonunified reality, it must adhere to unified form overall, reference by similarity, and emotional interpretation.

Similarity: A decorative shape or texture is usually extracted (or we might say abstracted) from some segment of actual experience, a surface texture, an image fragment, or a geometric pattern experienced in other contexts. The author of the visual should therefore take care that the associations evoked by border or background is consistent with the overall message of the visual. For example, Kostelnick and Hassett (2003) note that "an ornate border around a professional license might bolster its ethos, but the same border around a page of instructions would probably undermine it" (p. 103). Problems arise with any inconsistency between experienced associations and the message of the visual as a whole, as in Figure 12.

Other than as a joke, it would be absurd to send an employee a pink slip decorated with a floral border. In much the same way, no emotionally aware person would think of describing an infant who has just died as having "kicked the bucket." Either case would be considered a severe mismatch of idiom (visual or verbal) with a sensitive topic. Comparable mismatches can and do occur, however, when visual designers fail to consider decorative choices in relationship to informational messages, as seen in an actual cdc.gov Web site, intended to educate children about the risk of rabies developing in their pets. Instead of forms supporting an appropriately serious tone, we find a cheerful collision of playful typefaces, word-finding games, and brain dissection imagery. Images of the

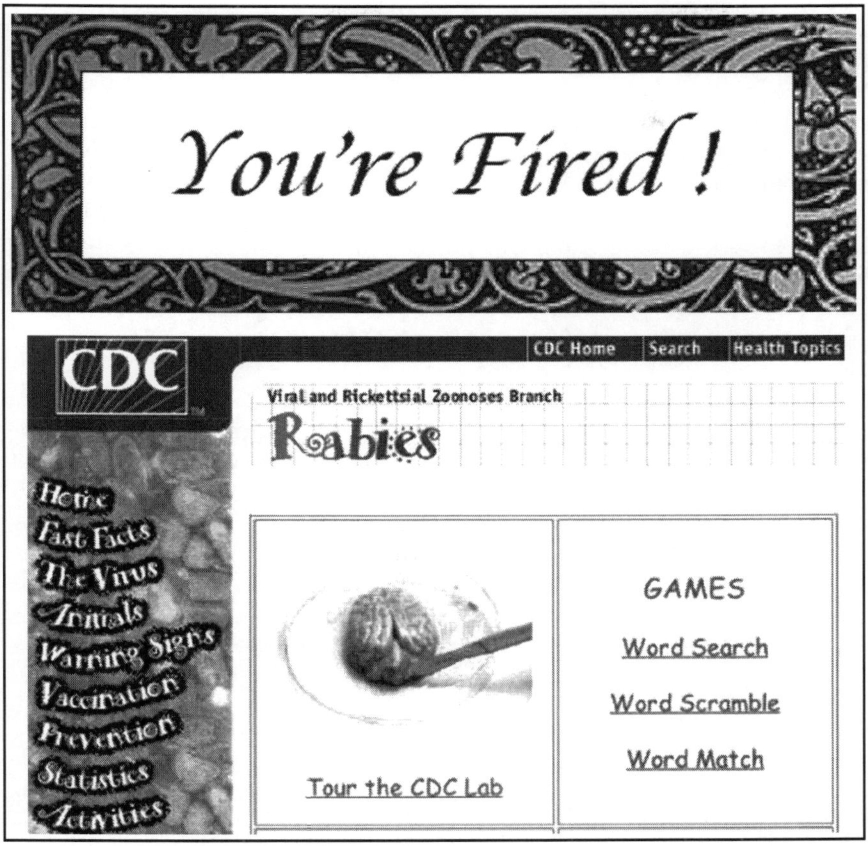

Figure 12. Problems matching decorative form to message.

Unity of feeling is necessary between decorative background and informative language, but is noticeably absent in the above examples, resulting in a feeling of clash.
http://www.cdc.gov/ncidod/dvrd/kidsrabies/Activities/cdclab.htm (accessed September 18, 2008).

rabies bacteria are transformed in a rainbow-colored ("fun") decorative border, further disrupting the unity of form and feeling.

Feeling: Along with specific, culture-based associations (e.g., khaki green associated with the U.S. military, white associated with weddings in European culture, but red in Asian culture, etc.), we can identify other interpretive associations that apparently arise from generic human perceptions and experiences.

In very general terms, we can define a range of feeling that a particular decorative form is likely to evoke, in the absence of specific cultural associations that otherwise influence the form's meaning. Here again, a Peircean analysis can help us to organize those generic associations. Table 2 (recalling the discussion of Table 1) identifies each of Peirce's three basic categories (as described in the Introduction) as a basis for associating, by similarity, a decorative form with a particular range of feeling. These generic form/feeling correspondences have been validated by various experiments, with subjects from several cultural backgrounds, where subjects were asked to match forms and feelings in the absence of cultural-context cues and either positive or negative conceptual evaluation (Amare & Manning, 2009; Collier, 1996; Collier et al., 1976; Ou et al., 2004; Xin et al., 2004; Zentner, 2001).

Thus, a background or border that has formal variety (e.g., a floral motif) but lacks strong contrasts or strongly ordered patterns will tend to evoke Peircean Firstness, usually manifest as a sense of diversion, feelings associated with ideas of freedom, potential, fun, and/or opportunity. Regardless of culture, there is evidently a common human experience of variety as something pleasurable, desirable, and fun.

A background or border that has stark contrasts but little variety or a strongly ordered pattern (e.g., jagged irregular angles) will tend to evoke Peircean Secondness, usually manifest as agitated feelings associated with ideas of collision,

Table 2. Generic Human Associations of Form and Feeling[a]

Basic Form of decorative	**Referential Process** Peirce-Category Similarity	**Interpretation Type** Range of Feeling
VARIETY	Firstness: freedom, possibility, opportunity	AMUSEMENT
SHARP CONTRAST	Secondness: collision, action, danger	AGITATION
PATTERN	Thirdness: order, understanding, principle	FOCUS

[a]In addition to specific cultural associations, generic human experience connects formal variety with freedom and play (here generically labeled as AMUSEMENT), sharp formal contrast with a need to pay attention to possible survival issues, food, love, or danger (generically labeled as AGITATION), and regular formal patterns with predictability and security (FOCUS).

action, and danger. Regardless of culture, there is evidently a common human experience with pointy things as potentially dangerous things.

A background or border that has a regularly repeated pattern (e.g., regular geometric shapes) without much contrast or variety will tend to evoke Peircean Thirdness, usually manifest as resolved feelings associated with regular, predictable patterns; that is, order, understanding, and principle. Regardless of culture, there is a measure of security found in regularity and predictability.

Examples of these primary form/feeling connections are shown in Figure 13.

Other kinds of feelings can be achieved by composition of the primary forms. Variety and pattern without strong contrast can create a mixture of pleasure and focus, for example, peaceful, contented feelings, or in other words, the very opposite of agitation. Contrast and variety without a strong pattern can create a

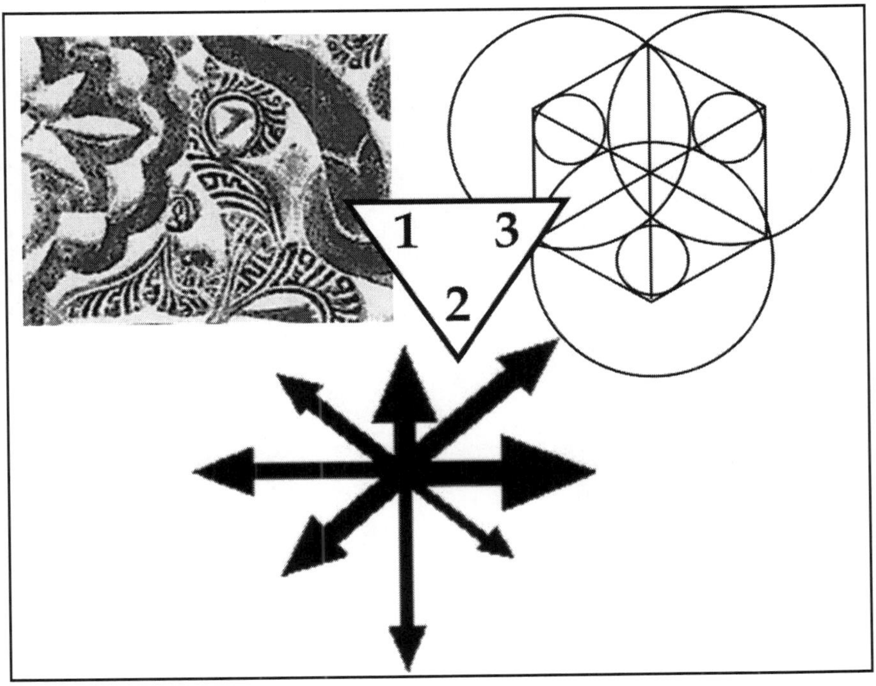

Figure 13. Primary form-feeling associations: (1) diversion, (2) agitation, and (3) resolution.

Generally speaking, formal variety feels like wandering amusement, sharp formal contrast feels like cause for agitation (positive or negative), and formal regularity feels like something focused or predictable.

feeling of interest or arousal or stimulation, that is, the very opposite of organized focus. Highly contrastive patterns without much variety can create a sense of unease or concern, for example, the very opposite of pleasure. Such effects are illustrated in Figure 14.

In discussing these general trends in form-feeling interpretation, we would emphasize that the feelings created by any decorative choice will be somewhat vague, and even if similar for most people, rarely exactly the same for any two individuals. This is due to the very nature of Peircean Firstness, diverse possibility as opposed to actual, specific fact.

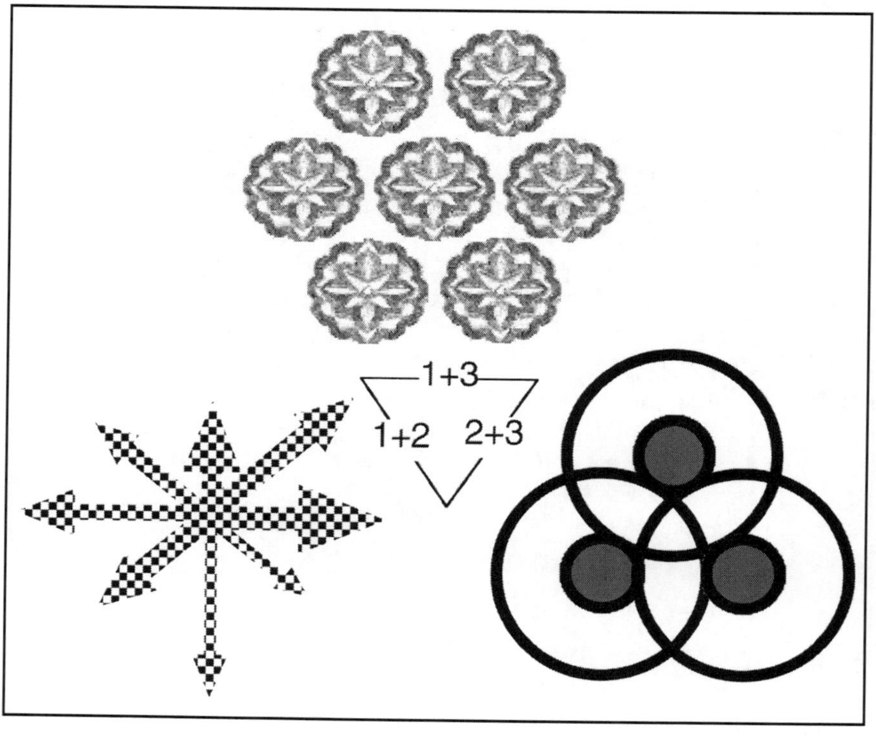

Figure 14. Composed form-feeling associations: (1+3) peace, (1+2) arousal, and (2+3) concern.

Mixing variety and pattern can create a feeling somewhere between diversion and resolution (i.e., calm or peace). Mixing variety and contrast can create a feeling between diversion and agitation (i.e., excitement or arousal). Mixing contrast and pattern can create a feeling between agitation and resolution (i.e., caution or concern).

Also, specific cultural associations can trump or overlay any generic perceptual associations based on variety, contrast, or pattern. Extreme varieties of form and color, for example, were typical decorative motifs in the United States in the late 1960s and early 1970s. For many, these motifs may still evoke feelings of fun and innocent play. Others may be reminded instead of the radical politics and drug culture of that era and therefore experience more negative feelings. The key point in Peircean analysis is that positive or negative evaluation is a kind of information and cannot actually be evoked by the decorative form alone: positive/negative evaluation must be established by some wider context.

Color and Typefaces

Decorative shapes and textures are, by their nature, usually distinct from nondecorative visual indicators and information. Colors and typefaces, by their nature, often are merged with nondecorative visuals, indicators, and information.

We often see differently colored arrows, bullets, or tags, and of course informative text must be presented in one typeface or another, and in one color or another, even if that color is most generically black on white, or white on black. Even so, we can identify the decorative aspects of color or typeface choice, and separate their decorative aspects from any indicative or informative aspect of a visual, simply by removing or changing the color or typeface and noting how feelings in the visual change while indicative and informative aspects remain the same (recall Figure 2 in comparison to Figure 1).

Apart from their typical merging with indicators and information, the essential principles governing decorative color and typeface choice are the same as those governing borders and backgrounds. Successful color and typeface choices must likewise create a sense of *unity*, a coherent meaning based on *similarity*, either to cultural associations or generic perceptual associations, and their decorative meaning will consist of a range of *feeling* related to either amusement, agitation, focus, or some intermediate combination of these primaries.

Unity: Decorative color and typeface choices likewise create an expectation of unified form, meaning that whatever typeface choice or color scheme is set out in the visual should be maintained throughout the whole. More than two or three typeface styles on the same page (not specifically creating an indicative focal point) instead create a problematic feeling of clash.

Color schemes may consist of colors in the same range (green-yellow-orange, or orange-red-magenta, or magenta-purple-blue, etc.), or contrasting pairs (yellow-red, red-blue, blue-yellow), or even sharply contrasting pairs (red-green, purple-yellow, or orange-blue), but as with decorative shapes and textures, whatever color motif is established in one corner of the visual should be somehow connected to the whole. To do otherwise will create a feeling of clash (Figure 15), equivalent to the border and background clashes illustrated in Figure 11.

THE *quick* **brown FOX**
jumped over
the lazy **DOG...**

Figure 15. Interrupted, inconsistent, and therefore problematic typeface and color schemes.

Typeface and color choices lacking perceptual unity will likewise create feelings of clash. Compare Figures 5, 7a, and 11.

Similarity: Interpretations of colors or typefaces likewise derive from perceived similarities to prior experience. Some typefaces strike us as more professional or business-like because we've encountered them in such settings. A certain shade of yellow may please us or make us feel somewhat queasy, depending on whether we associate it with a favorite flavor of ice cream or with illness/infection. The author of the visual should therefore likewise take care that the associations evoked by color and typeface choices are consistent with the overall message of the visual. Problems likewise arise with any inconsistency between experienced associations and the message of the visual as a whole, as in Figure 16.

As in Figure 12, it is difficult to reconcile the variety of color and form (i.e., the playful-diversion feeling) in the header typeface in Figure 16 with the gravity of its message.

Feeling: Along with specific, culture-based associations (e.g., red, white, and blue color associated with U.S. politics or gothic typeface associated with medieval culture), other interpretive associations of color and typeface may arise from generic human perceptions and experiences. As with border/background forms, we can define a range of feeling that a particular color or typeface shape is likely to evoke, in the absence of specific cultural associations that otherwise influence the form's associated meaning. Here again, a Peircean analysis can help us to organize some useful generalizations. Table 3 and Table 4 (recalling the discussion of Table 1 and Table 2) identify each of Peirce's three basic categories as a basis for associating, by similarity, a color range (Table 3) or a general typeface style (Table 4) with a particular range of feeling.

Unless trumped by specific cultural or personal associations (e.g., pale yellow to illness, etc.), color choices in the yellow range, like highly varied

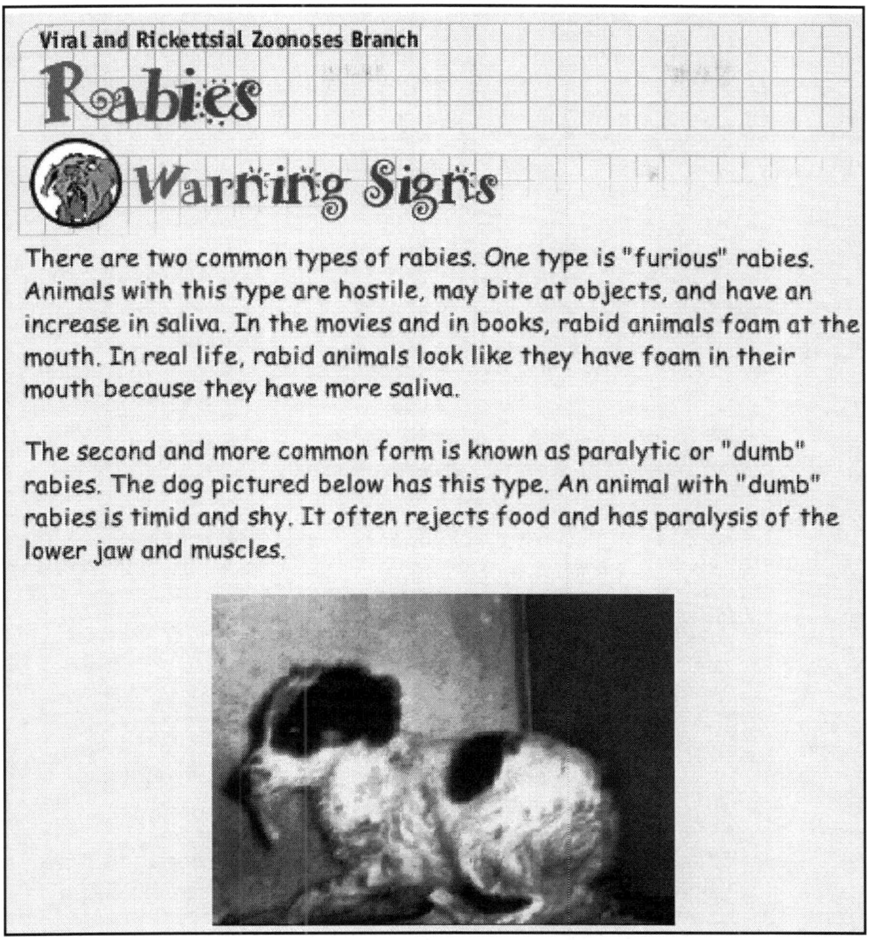

Viral and Rickettsial Zoonoses Branch

Rabies

Warning Signs

There are two common types of rabies. One type is "furious" rabies. Animals with this type are hostile, may bite at objects, and have an increase in saliva. In the movies and in books, rabid animals foam at the mouth. In real life, rabid animals look like they have foam in their mouth because they have more saliva.

The second and more common form is known as paralytic or "dumb" rabies. The dog pictured below has this type. An animal with "dumb" rabies is timid and shy. It often rejects food and has paralysis of the lower jaw and muscles.

Figure 16. Further problems matching color and typeface choices to message.

Header typeface and color variety chosen by visual designers again inconsistent with the tone of the information. Compare Figure 12.

Table 3. Generic Associations of Color and Feeling[a]

Basic Form Color Range	**Referential Process** Peirce-Category Similarity	**Interpretation Type** Range of Feeling
YELLOW green <--> gold	Firstness: freedom, possibility, opportunity	AMUSEMENT
RED orange <--> magenta	Secondness: collision, action, danger	AGITATION
BLUE purple <--> aqua	Thirdness: order, understanding, principle	FOCUS

[a]In addition to specific culture associations, generic human experience may connect color in the orange-yellow-green range with freedom and play (here generically labeled as AMUSEMENT), color in the orange-red-purple range with a need to pay attention to possible danger (generically labeled as AGITATION), and color in the purple-blue-green range with predictability and security (FOCUS).

Table 4. Generic Associations of Typeface Styles and Feeling[a]

Basic Form Color Range	**Referential Process** Peirce-Category Similarity	**Interpretation Type** Range of Feeling
CURVED-ARABESQUE *italic <--> calligraphic*	Firstness: freedom, possibility, opportunity	AMUSEMENT or "Literary Elegance"
HIGH CONTRAST **bold <--> BLOCK**	Secondness: collision, action, danger	AGITATION or "Forceful and Concrete"
PATTERNED regular <--> serif	Thirdness: order, understanding, principle	FOCUS or "resolved, comfortable"

[a]Though somewhat muted, the feelings evoked by typeface choices follow the same Peircean principles of form-feeling association discussed in relation to form and color.

border/background patterns, tend to evoke pleasurable feelings associated with ideas of freedom, potential, and opportunity. This is perhaps due to universal human experience with the color of early morning sunlight, generally associated with a new day in pleasant weather, typically full of potentials and possibilities.

Color in the red range has an effect similar to a border that has stark, sharp contrasts and will tend to evoke agitated feelings that can be read positively as motivated action or negatively as collision or danger. This is probably due to the common instinctive reaction to the sight of blood: survival requiring an immediate response to the color red in that case, a negative reaction if the blood is yours, a positive reaction if the blood is from a prey animal or causing the high color of a lover's cheeks.

Even so, the basic agitating properties of red can easily be blended with patterned form, for instance the Target store logo of concentric circles and the net effect is then a feeling of activity (the contribution of red) plus focus (the contribution of the pattern), and this conforms with the meaning of "Target," and thus actively draws viewers to the store.

Color in the blue range tends to evoke focused or organized feelings similar to those associated with regular, predictable patterns; that is, order, understanding, and principle. This is perhaps due to common human association of the color blue with distant mountains, clear sky, and calm sea, with constancy, predictability, perspective, etc.

As with borders and backgrounds, other kinds of feelings can be achieved by a composition of one primary color with a different primary, or by combining the effects of a color scheme with the effects of variety, contrast, or pattern summarized in Table 2. For example, yellow-range colors combined with formal variety can reinforce a feeling of amusement, pleasure, freedom, etc. On the other hand, a yellow background combined with formal variety and red foreground can create a feeling of arousal or excitement, halfway between pleasure and raw agitation. Highly contrastive patterns in the blue range, or red mixed with blue, can create a sense of unease or concern, halfway between agitation and resolution.

Such compositions are typically found in movie posters, the colors chosen having some correlation with the overall feel of the film advertised, as in Figure 17.

Mixing variety and pattern can create a feeling somewhere between diversion and resolution (i.e., calm or peace). Mixing variety and contrast can create a feeling between diversion and agitation (i.e., excitement or arousal). Mixing contrast and pattern can create a feeling between agitation and resolution (i.e., caution or concern).

As always, the feelings created by any color/form choice will be somewhat vague, and cultural associations can trump or overlay any generic perceptual associations based on variety/yellow, contrast/red, or pattern/blue.

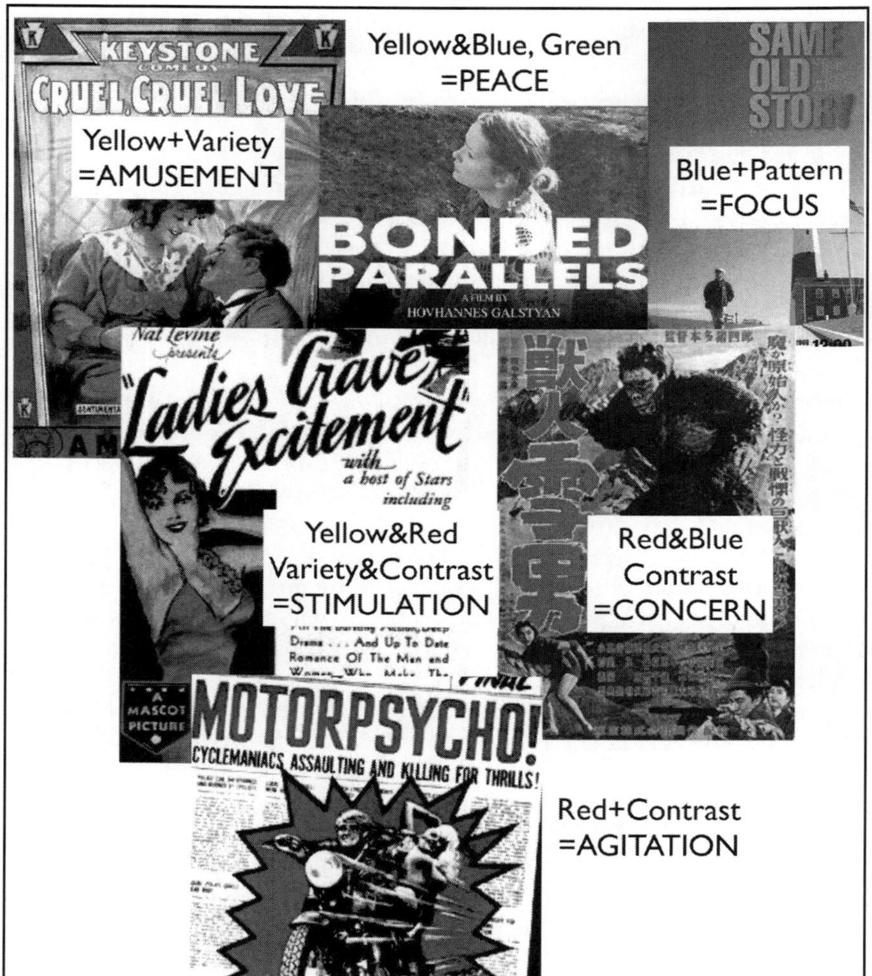

Figure 17. Composed color/form-feeling associations: peace (1+3), stimulation (1+2), uneasy concern (2+3).

Color-form compositions are typically found in movie posters, the colors chosen having some correlation with the overall feel of the film advertised.

Similarly, because typeface forms must accommodate the conventionalized informative properties of printed letters, the generic emotional effects of typefaces, compared with those of color or raw form, are somewhat muted. Nevertheless, we find that varied, arabesque forms of italic or calligraphic typefaces evoke a sense of fun or free "friendly" amusement but often in a more muted sense of artistry, literary amusement, or elegance (Mackiewicz & Moeller, 2004, p. 308).

High-contrast bold or block typeface styles are rarely felt as risky or agitating, but will be felt as more forceful or concrete. Used to excess, a bold or block style can fatigue viewers, even to the point WHERE THEY FEEL THEY'RE BEING SHOUTED AT, and in that sense, a feeling of collision or physical irritation is still implicit in a high-contrast typeface style. In more muted ways, high contrast in the uppercase-lowercase features of a typeface, or thickest-to-thinnest line, or width-to-height can all create feelings of agitation or discomfort that inhibit readability (Mackiewicz, 2006, p. 74).

Typefaces that exhibit a more regular or consistent pattern, with serifs or without, tend to create the feeling typical of all regularly patterned forms, that is, a resolved sense of continuity and order, which typically translates into a sense of professionalism (Mackiewicz & Moeller, 2004, p. 308).

Here again, we need to emphasize that deliberate cultural-iconic associations can override these basic typeface effects. There are display typefaces, for example, in the style of dripping blood, which can recover the full emotional force of danger. Gothic-style typefaces can trigger either positive religious/historical connotations or negative horror-genre connotations, depending on context, and so on.

CONCLUSION: TOWARD A DECORATIVE ETHIC

As discussed in the Introduction, ethical action is here defined as action directed toward sustainable, attainable goals, goals acceptable to both the actor and all those who are acted upon. Visual information design is just one specific domain where this general model of ethics applies: the creator of a visual acts upon viewers with a visual to accomplish whatever goal that all are collectively pursuing, assuming that the goal is attainable by the means chosen and sustainable over the long term (Manning & Amare, 2006).

- This definition of ethical action covers the obvious requirement that a visual be accurate in its information. Whether or not the visual's author was being deliberately deceptive, an inaccurate visual fails to ethically meet the goal of the community collectively: to know the truth.
- This definition of ethical action covers what should be the equally obvious requirement that a visual not harm any of its viewers or the wider community

acted upon by what the viewers of the visual do. Conspiracies of smaller groups to commit fraud, tyranny, ecological destruction, or genocide upon larger groups fail the definition of ethical action, primarily because the goals of conspirators don't match the goals of everyone affected, and also because conspiratorial goals are "false" in the sense that they can't be attained or sustained over the long term: fraud tends to be discovered, tyranny resisted and overthrown, while ecological destruction and genocide alike ultimately turn out to be elaborate forms of suicide.

This definition of ethical action also covers less obvious requirements:

- That visuals be as clear as possible in their information, as clear as possible in what they indicate, and in what feelings they evoke; otherwise, the goal of viewers to understand what they're looking at is frustrated.
- That the strategies used to construct visuals be suited to actually attaining their apparent purposes, that decorative strategies, for example, not be used to try to convey information; otherwise, informative goals of both the visual creators and viewers, for that particular visual, are unsustainable.

The more obvious ethical requirements of accuracy and noninjury, as they apply to visual design, have been discussed elsewhere (Dragga, 1996; Dragga & Voss, 2001; Elliot & Lester, 2002). Our particular concerns in this book are the clarity and suitability of visual design strategies relative to specific goals. Our particular concerns in this chapter have been the clarity and suitability of decorative design strategies for various kinds of visuals.

In general terms, it is less than ethical to use decorative elements in uncontrolled, inconsistent ways, such that the feelings that they evoke are confused (recalling, for example, Figures 7a and 15).

It is also less than ethical to use decorative strategies to the point of distraction, when the apparent purpose of a visual is indicative or informative. When designers highlight in a visual their personal prowess with decorative elements of form, color, and imagery for an audience that is primarily seeking clear information, there is a mismatch of goals and an ethical breach (recalling for example, Figure 12, bottom).

Conversely, positive ethical value is manifest whenever a visual designer has obviously taken care to match decorative form to a specific emotional purpose (recalling, for example, Figure 4 of this chapter). Take, for instance, the decorative choice in the following example:

> For example, a hard-edged drawing . . . makes the building appear technical, precise, and objective—qualities appropriate for certain groups of readers like engineers, contractor, or clients. On the other hand, rendering the same object in freehand style . . . changes the tone immediately by decreasing its

> formality and making the image appear more accessible and nonthreatening, even though its design differs only slightly from the hard-edged drawing. As a result, freehand or cartoon-like figures (or clip art images) are frequently used in instructional materials where readers may perceive the task as difficult or threatening and may need motivation and assurance to complete it. In these situations, conventional stylistic cures meet the pragmatic need to express an appropriate tone. (Kostelnick & Hassett, 2003, p. 102)

In terms of a decorative ethic, the designer of the line drawings chose to use not-so-straight lines and blotchy shading to make the drawing look less technical. However, problems may arise if the audience is insulted by a juvenile drawing. This is why usability tests are important to conduct repeatedly in order to increase knowledge of audience needs.

To some degree, we can excuse decorative-strategy misfires when they are not deliberate, but only to the same degree that we excuse factual inaccuracy from someone who does not intend to deceive (Manning & Amare, 2006). There may not be guilt or criminal liability, but ethical responsibility still rests with the creator of the visual for both the accuracy and the effectiveness of the visual, even at a decorative, emotional level. We will see in the next chapter that many considerations governing decoratives in general also apply to images in general, that they are most effectively chosen for a specific emotional purposes not interfering with larger indicative and informative purposes.

CHAPTER 2

Images

We will here use the term *image* in a tightly restricted sense. Visuals in general are often referred to as "images," but we intend the term, in its technical sense, to designate visuals reflecting the actual physical appearance of objects. Images may be largely decorative, but they also inherently evoke certain indicative properties, in that, in Peircean terms, they are one step down, in the direction of indicative purpose as shown in Figure 2 and Figure 4 in the Introduction. Each image tends to provoke viewers to actively reference some specific object or kind of object in the domain of discourse. This passage by Sedlack, Shwom, and Keller (2006) denotes *image* (here, a picture) as one type of specifically decorative visual that references kinds of objects (in this case, healthy people) rather than specific individuals:

> In some cases, visuals need no textual explanation. Consider, for example, the annual report for a pharmaceutical company, which includes pictures of active, healthy people. The report is likely never to refer to these pictures explicitly. Instead, the role of the visual is to evoke a *feeling* in the reader. The picture implicitly says this company is committed to good health. (p. 65)

Because contemporary usage of the term *image* covers a wide range of visuals, we need to be specific about what images will refer to in the Peircean typology. Taking the various dictionary definitions of *image* (e.g., the first seven entries from Dictionary.com), we specify that definitions 1, 2, 4, and 6 below follow our technical definition, but definitions 3, 5, and 7 (boldface next page) are not sufficiently precise.

1. a physical likeness or representation of a person, animal, or thing, photographed, painted, sculptured, or otherwise made visible.

2. an optical counterpart or appearance of an object, as is produced by reflection from a mirror, refraction by a lens, or the passage of luminous rays through a small aperture and their reception on a surface.
3. **a mental representation; idea; conception.**
4. *Psychology*. a mental representation of something previously perceived, in the absence of the original stimulus.
5. **form; appearance; semblance: *We are all created in God's image.***
6. counterpart; copy: *That child is the image of his mother.*
7. **a symbol; emblem.**

In other words, not all representations based on resemblance are images in the precise sense of physical reflection or direct perceptual recall, as indicated by definitions 1, 2, 4, and 6 given above. Representations based on any kind of resemblance were defined by Peirce as *icons*, and he was careful to differentiate three types: *qualisigns*, *sinsigns*, and *legisigns* (CP 2.244–246, 8.334).

We will see that there are important practical distinctions to be made between pure decorative forms or *qualisigns* (1), images or *sinsigns* (2), and diagrammatic forms or *legisigns* (3), as shown in Figure 1.

As suggested by most of the dictionary definitions given, it is physical reflection of objects or perceptions that differentiates images, in this technical sense, from nonimage decorative forms (Figure 1, top left) and on the other hand, from diagrams (which include cartoons and maps as subtypes, Figure 1, top right), and also differentiates images from true symbols (such as hieroglyphics), which come to reference abstract word concepts more than physical objects that the hieroglyphic forms originally resembled.

Nonimage decoratives consist of raw qualities abstracted from perception (color, shape, texture, etc.) but not tied to the appearance of any particular physical objects. Cartoons, maps, and diagrams in general consist of relationships abstracted from knowledge and are distinctly remote from either the perception or the physical attributes of any actual thing. Maps, an exemplary kind of diagram, little resemble the actual physical terrain—regions are colored, borders are marked, etc. in ways that are not visible when looking at the actual landscape.

The obvious disjunction between diagrammatic representations and physical appearance typical of images is occasionally the subject of parody, as for instance in this dialogue between Twain's Huck Finn and Tom Sawyer.

> "Because if we was going so fast we ought to be past Illinois, oughtn't we?"
> "Certainly."
> "Well, we ain't."
> "What's the reason we ain't?"
> "I know by the color. We're right over Illinois yet. And you can see for yourself that Indiana ain't in sight."
> "I wonder what's the matter with you, Huck. You know by the color?"
> "Yes—of course I do."

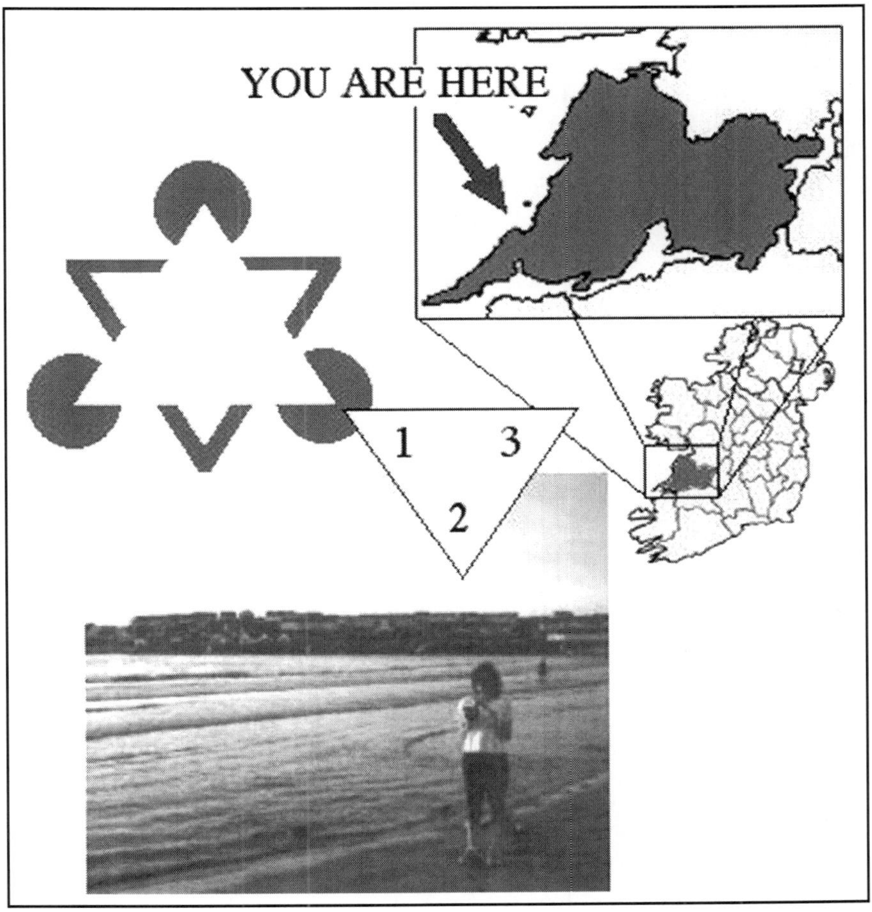

Figure 1. Pure decoratives vs. images vs. diagrams.

All representations based on resemblance are icons, including (1) abstract forms = Peircean qualisigns, (2) images = Peircean sinsigns, and (3) maps and other diagrams = Peircean legisigns. Images in the technical sense (sinsigns) are only those icons based on physical resemblance or reflection.

"What's the color got to do with it?"
"It's got everything to do with it. Illinois is green, Indiana is pink. You show me any pink down here, if you can. No, sir; it's green."
"Indiana pink? Why, what a lie!"
"It ain't no lie; I've seen it on the map, and it's pink.
(Twain, as quoted in Barton & Barton, 2004, p. 234)

The take-home message of Twain's satire is that we can't expect viewers to automatically extract specific kinds of information (like state boundaries) from direct physical appearances alone. Looking at the bank of a river (or a photograph of it) can't tell you directly what state it borders on, how many fish of what kind are in the river, or what population of beavers lives along its shore. As a consequence, images by themselves are not reliably informative. Images do contain potential information, just as orange color and a dancing form might potentially indicate a fire, as well as an indefinite number of other orange dancing things. So too, images potentially could *indicate* information to an expert, but that expert would in fact be matching patterns of prior knowledge to the image that a nonexpert without that knowledge would never see in the image alone.

Recall now the Peircean grammar discussed in the introduction: Firstness-feeling and Secondness-indication do not automatically imply Thirdness-information. However, the Secondness-objects essentially indicated by images do imply Firstness-feelings. Therefore, images indicating physical objects do gravitate toward decorative uses (because Secondness implies Firstness). It is utterly common for images to be used to convey feelings associated with the objects that the images resemble (as for example, in a decorative poster featuring a photograph of a cute kitten), but images can also be used as indicators, actively focusing reader attention on real and specific objects that the images resemble (as in a poster of a lost cat featuring a photo of the specific lost cat, which aids in a search for the animal).

Unlike more-abstract decoratives (border forms, background colors, textures, or shapes), images more directly evoke the physical appearance of specific objects.

In this era of digital photography, images are the easy visual-design choice. Most word-processing programs (e.g., Microsoft Word) allow you to "drag and drop" image files into a text with single move-click operation. The easy choice has quickly become the standard choice, even in technical communication genres. For example, Ann Jennings in "The Sales Aspect of Engineering Presentations" (2007) encourages engineers to use images (such as a picture of a steering wheel or a van) to showcase the feeling of the highly engineered steering. Similarly, Michael Alley's popular *The Craft of Scientific Presentations* (2003) offers exemplary PowerPoint models with far more images than diagrams in the slide deck, even when the scientific slides have a purportedly informative purpose.

The easy design choice is often a dangerous one, however. Images are exactly like all other decorative forms in that they cannot (by themselves) directly convey any definite information. A satellite photograph does not, by itself, identify roads, buildings, land forms, etc. Recall that we use the term *information* in the technical sense of specific truth-valued propositions: *This is I-80; this is where the Mississippi borders Illinois,* etc.

An image by itself can neither affirm or deny a proposition since the physical reality reflected by the image, contrary to Huck's assumption, doesn't come color-coded or otherwise labeled, but maps implicitly assert a number of identifications and relationships.

The common perception that "a picture is worth a thousand words," besides being a misquotation from the start (Lester, 2004) is only an illusion because an image will mean many different things to different people, depending on context, and the fact that no specific words can reliably be derived from an image. There are instead a thousand (or more) different words that might be attached to an image, many of them not compatible in meaning. The message of images is always open and potential rather than fixed. As shown in Figure 2, the meaning of an image is easily skewed by the verbal caption that accompanies it.

The multiple possible interpretations of any given image can be used effectively, to affect a wide audience with a strong but nonspecific emotional message, as illustrated by the famous Apollo 8 Earthrise image (Figure 2).

Of greater concern, however, are the information-content weaknesses inherent in images, due again to multiple interpretations. Images are the easy design choice, but they are not an effective choice, by themselves, when authors need to convey very specific informational messages. An unfortunately typical scenario is described by Walsh:

> Speakers were assigned to give thumbnail sketches of various periods of art, each presentation to be followed by a walking tour to observe actual examples of the period being studied. Unfortunately, some of the speakers illustrated their lectures only with photographic slides, hoping that the audience would be able to identify the relevant details. One speaker used literally hundreds of slides to illustrate Baroque art without ever managing to explain exactly how one might identify a Baroque church, as opposed to, say, one from the Renaissance or Early Christian periods. Two and a half hours and two hundred detailed photographs resulted in total confusion, but a half-dozen good line drawings showing what detail is relevant to each period would have done the trick in fifteen minutes. (1998, pp. 209–210)

In situations such as Walsh describes, relatively few images can more efficiently be used as illustrative examples, if the conceptual points are first made clear by simplified line drawings (i.e., diagrams).

Conversely, even without clarifying diagrams, advertising or political persuasion can make effective use of images since feeling-creation is a key part of their persuasive task, especially in cases where detailed information about a product or social agenda is not as critical to the purchase or social policy decision.

As shown in Figure 3and 4, images can be added to persuasion with good results, especially if used deliberately and ethically. As was noted in the previous chapter, well-used decoratives, including images, can help put readers at ease as more informative kinds of visuals are also presented—diagrams, graphs, tables, etc. However, informative visuals typically are not enhanced by direct integration

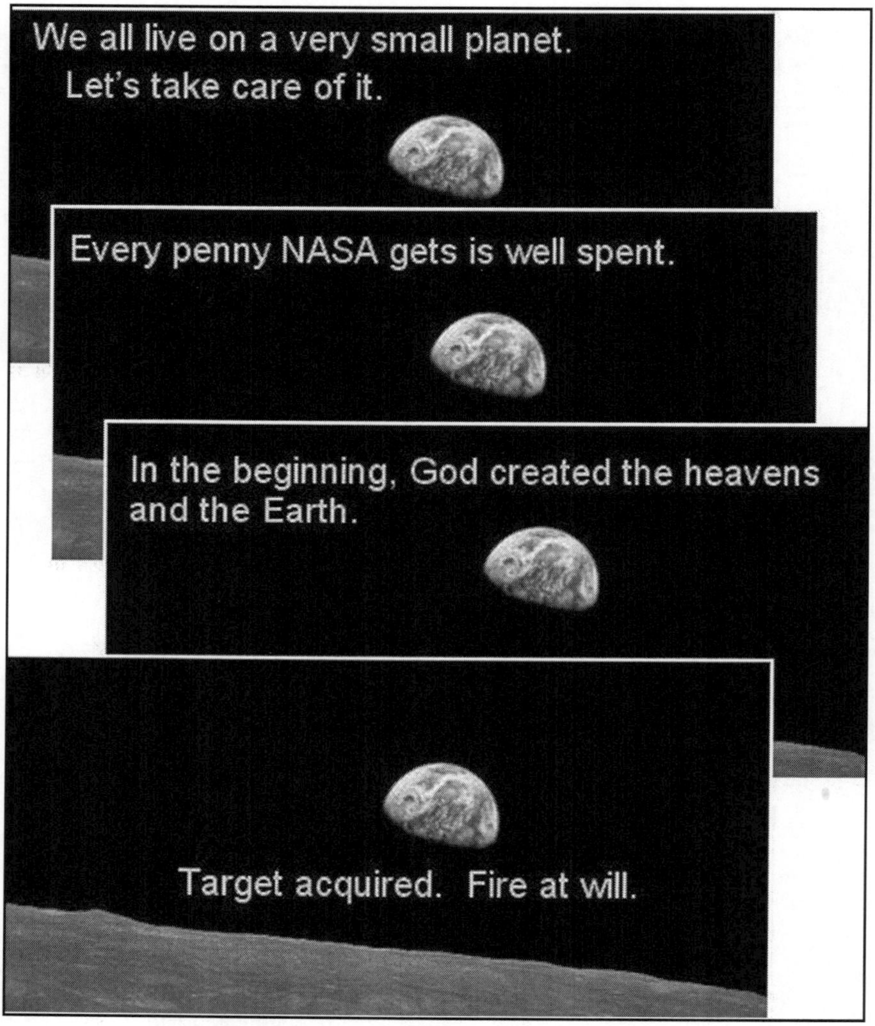

Figure 2. One image with multiple captions equals multiple meanings.

A single image can be made to distinctly convey messages in support of ecology, technology, theology, or dystopian anxiety, depending on the context.
Photo: Astronaut Bill Anders (Apollo 8) of Earthrise over the moon (nasa.gov)

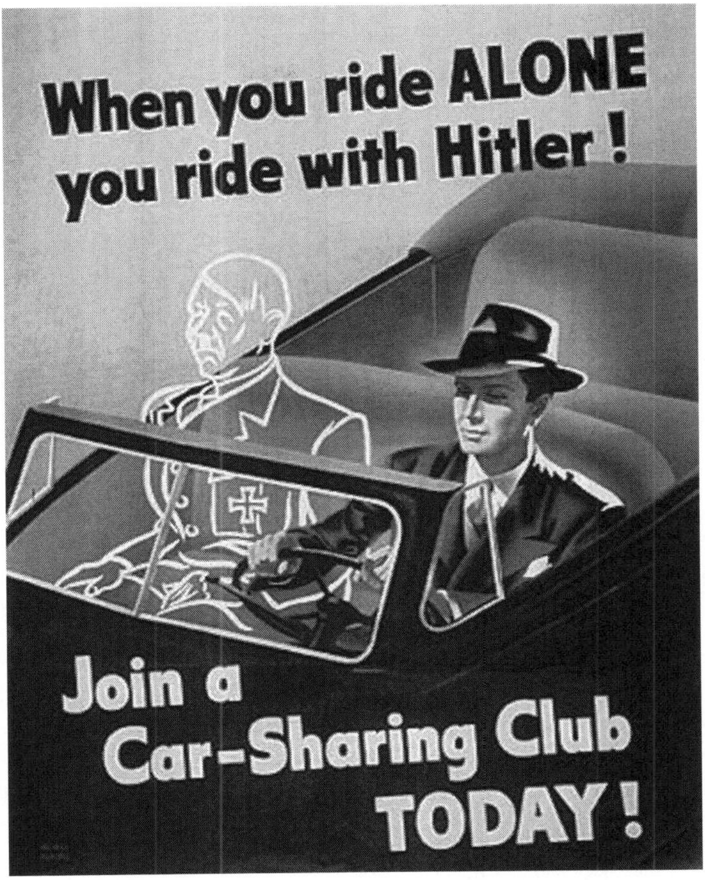

Figure 3. Public service message enhanced by visual.
http://www.archives.gov/exhibits/powers_of_persuasion/use_it_up/images_html/ride_with_hitler.html

of images. Images serve two essential functions: to create a proper mood or feeling, and/or to indicate particular kinds of objects, as well as specific objects, people, or events.

Neither of these functions is particularly helpful, for example, in Figure 5a, which informs viewers about a local trash pickup schedule. Presumably, the viewers have already identified their trash bins, and there is no useful feeling or emotion to be evoked by images of trash bins in this context. Viewers do need to know which kind of bin (indicated by the bin lid color) is emptied in which week, but again the whole image of the trash bin is unnecessary for that indicative purpose. Likewise, the images in Figure 5b are neither evoking

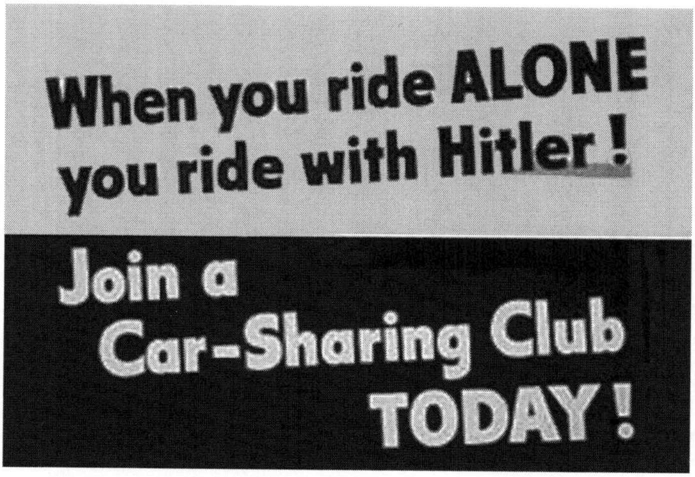

Figure 4. Message somewhat drab without image elements.

an essential emotion, nor are they identifying people who are important to the task of obtaining a visa from the U.S. government. The actual purpose of the Web page is to forewarn viewers that some steps in the visa process have changed and direct them to a link that leads to the necessary forms and procedures. The images here suggest that there is a certain camaraderie shared by all those caught up in the same red tape of a visa application, but these representative faces do not contribute to the essential instructional purpose of the Web page and therefore constitute a likely distraction. Extra images, like unnecessary wordiness in the text itself, can actually make the task feel more complex than it is.

An efficiency-conscious editor could easily remove these images and still serve the essential informative purposes of either of these visuals, as shown in Figure 6. Such editing removes the potential for distraction and increases the feeling of focus in the visuals.

Information can suffer when image elements are overused to the point of distraction, most especially if they are used as lazy, possibly even deliberate substitutes for more appropriate indicative and informative elements, indicators and information that are more useful to viewers but that require more effort for the visual creator to integrate with decorative purposes more easily served. Scott Adams (2011) has deftly satirized this kind of deliberate image-for-information substitution:

> BOSS: Dilbert, work with Gustav to get our new product explained on our website.
>
> GUSTAV: Here's what I have so far.

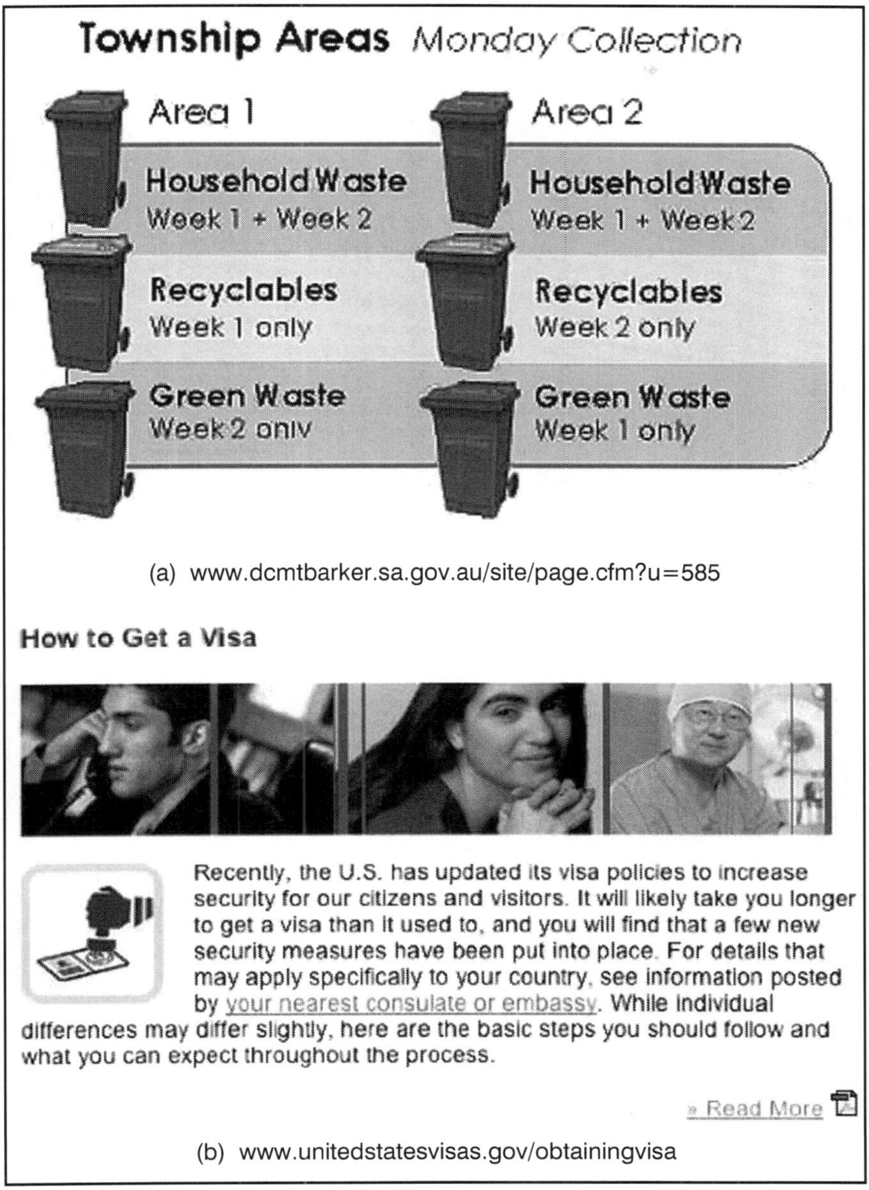

(a) www.dcmtbarker.sa.gov.au/site/page.cfm?u=585

(b) www.unitedstatesvisas.gov/obtainingvisa

Figure 5. Image-cluttered informative graphics.

Images not directly serving the informative purposes of a visual instead are likely to distract viewers and/or make the visual feel unfocused.

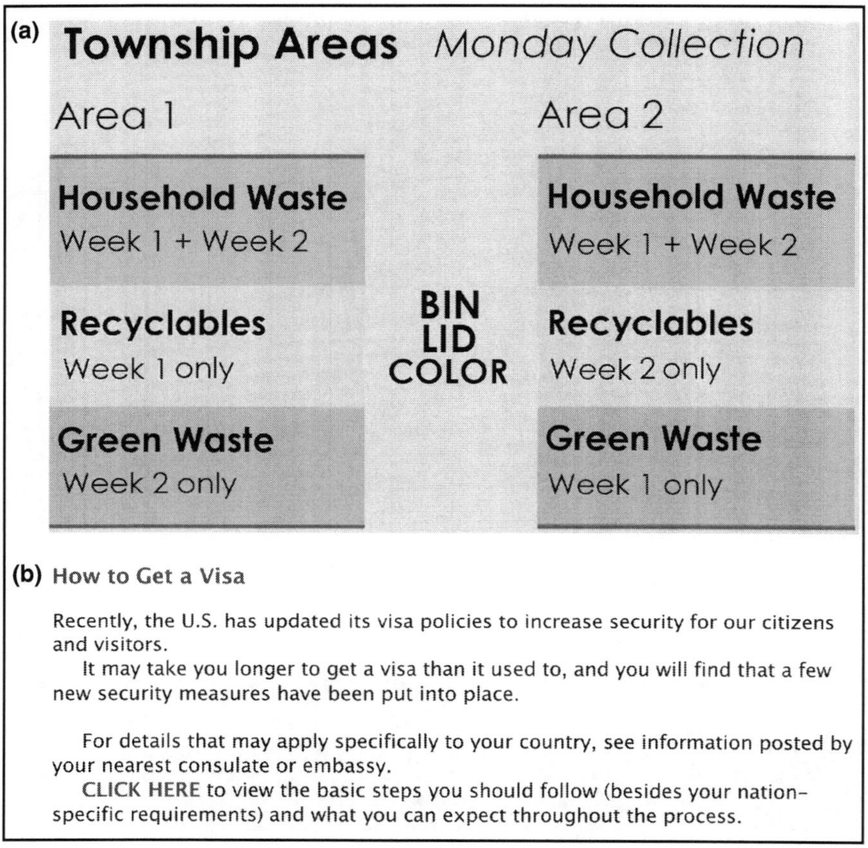

Figure 6. Image-removed informative graphics.

> DILBERT: It's awful.
>
> GUSTAV: Excuse me?
>
> DILBERT: There's no information. It's all images and annoying music. People make buying decisions based on what they read. This gives them nothing.
>
> GUSTAV: If I clutter the design with useful information, it will look ugly and I won't be able to use it in my portfolio.

More-obvious cases of image-based deception occur when advertised images or the images on product packaging differ from the actual product (Figure 7a). More subtle in their deception are images (and other decorative forms) that make

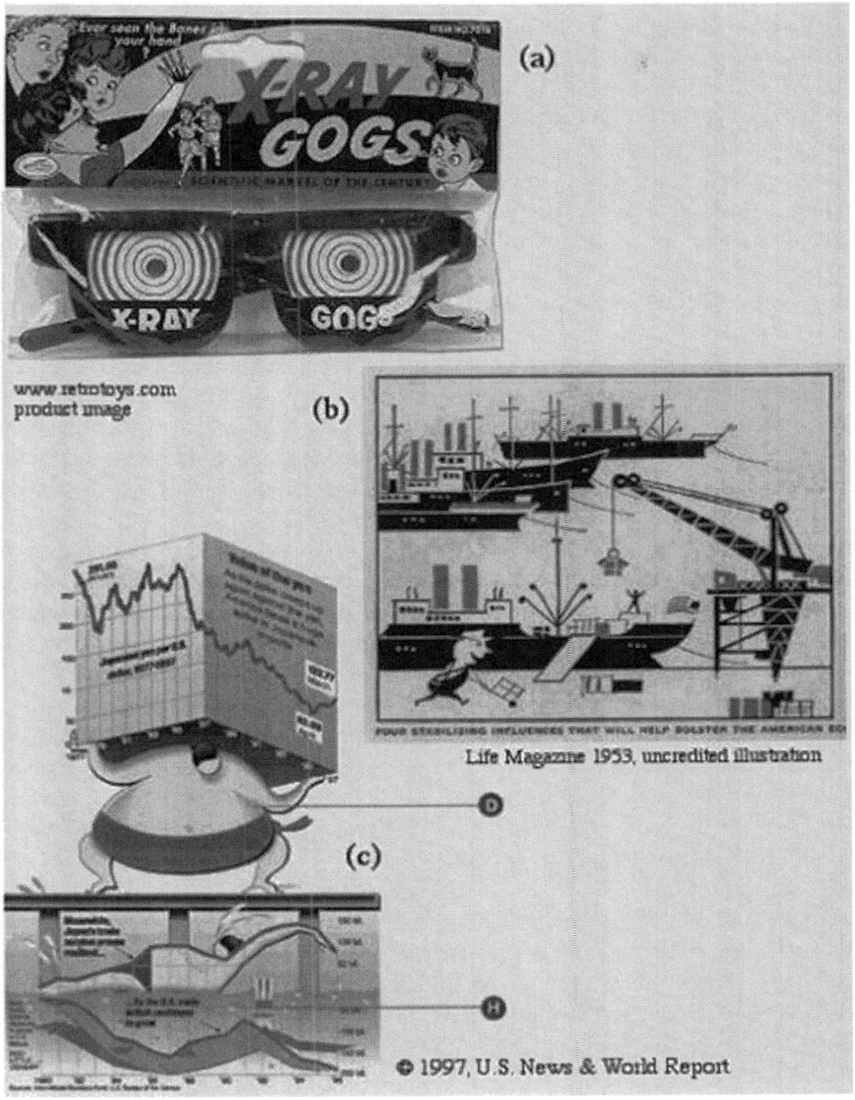

Figure 7. Images distorting, substituting for, or distracting from actual information.
http://www.macworld.com/article/2270/2001/06/howtodata.html

a visual seem more useful, more informative than it actually is. The caption for Figure 7b, for example, identifies "four stabilizing influences" for the U.S. economy, but none are clearly identified in the illustration.

Conversely, the essential information may be present, but images mixed with the data can distract viewers and keep them noticing important points—a different kind of deception—but still ethically suspect (Figure 7c).

Figure 7c is particularly problematic precisely because its imagery does not directly obscure or distort any of the data displayed in the embedded line graphs. Other analysts have actually praised the graphic on these grounds (e.g., Abes, 2001), but in doing so, they confuse the unquestionably appealing properties of the visual with its actual readability. In particular, the sumo wrestler imagery merged with the head-above-water imagery (a mixed metaphor at best) biases the reading of the data toward the big-picture conclusion (in the most literal sense of *big picture*) that the Japanese economy is "beating" the U.S. economy. The imagery can therefore distract viewers and keep them from noticing potentially crucial details in the data, for example, the most recent downturns in the yen exchange rate and the Japanese trade surplus (Figure 8).

Ethical problems would definitely arise if the visual in 7c were being used to entice investment in Japanese currency or securities at the precise time when those investments were in fact no longer going to show significant growth. In that scenario, the triumphant sumo imagery is identical in effect to the famously deceptive marketing of X-ray glasses, where the promotional imagery will not align with actual performance.

In this chapter, we will discuss images within the explanatory framework of Peirce's visual types, the legitimate use of images as decoratives and as indicatives but also their overuse and unethical use in persuasive and informative texts.

IMAGE DESIGN—PEIRCEAN PARAMETERS

Applied to the form, reference, and interpretation of visuals, Peirce's basic categories can help us to better understand the design parameters of each visual type, its essential form, its range of application, and its limitations. Recall that each visual type has three key parameters:

- its basic form—whether the visual is unified or has distinct phases;
- its referential process—how the visual acquires its meaning;
- its interpretation type—what kind of meaning the visual form typically acquires.

Basic Form

Like pure decorative forms, an image (in our technical sense) requires a form that creates *qualities of unity*. This means that images, like all decoratives, should

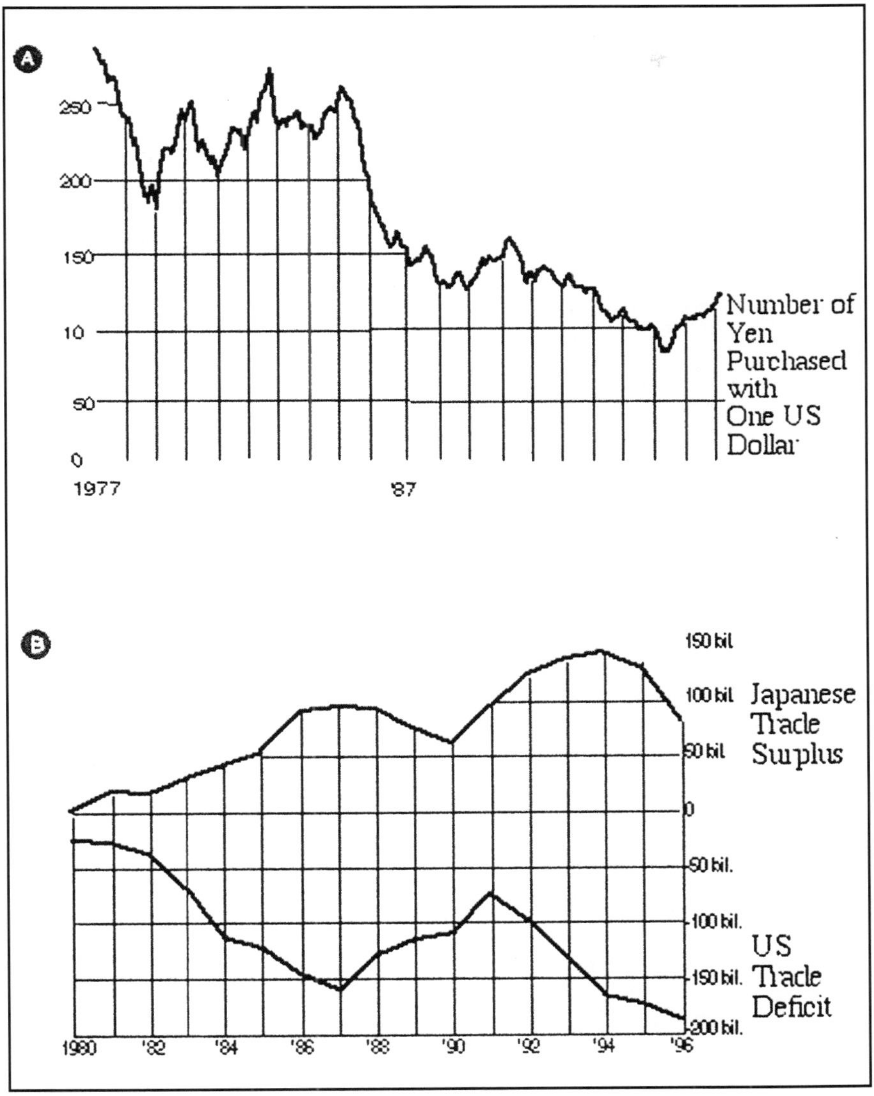

Figure 8. Data from Figure 7c without image distraction.
Cathy Abes, "Bring Data to Life: Art and Information Can Complement Each Other," Macworld.com, June 2, 2001.

Without the bias of imagery toward a "big picture" interpretation, viewers can more easily notice small but significant details in the data.

generally follow gestalt principles such as balance, symmetry, and continuity. All elements in a single well-formed image constitute a perception with implicit unity. Consider cases where an original image is noticeably marred in some way, the sense of unity disrupted; then, it becomes something other than a mere image. For example, a portrait that has been vandalized by being torn in half or by having a mustache crudely drawn on it (Figure 9) is no longer taken as a reflection of the individual but rather as some kind of negative comment on the individual or the artwork (becoming a social signal, which is Peircean Type VII [code 2-2-3] rather than a mere image, type II [code 1-1-2]. Social-ritual signals will be discussed further in chapter 5).

Referential Process

Like pure decorative forms, an image gets referred to its interpretation through *qualities of similarity*. This means that an image, like decoratives generally, can only be understood through its similarity to other perceptions in viewer memory. Image reference results from perceptual associations between the form of the current image and a viewer's actual perceptual experience. The effect of an image, more often than not, depends to a large degree on viewers' specific prior experiences.

Generally, image-driven design choices and decorative design choices can therefore misfire if the author misjudges the associations that viewers will make with images or decorative forms presented. Figure 10, an advertisement acceptable in its day, would now be felt to be unacceptably sexist.

Also of note in Figure 10, the image of the "girl" was intended in its day to reference a young, stylish individual, but now the image would seem to reference someone older and somewhat dowdy due to the associations we would now make with that (grandmotherly) style of clothing.

Interpretation Type

The interpretive *physical-object connection* between images and actual objects marks the essential distinction between images and other more purely decorative forms, on one hand, and diagrams on the other, recalling Figure 1. Images directly reflect objects, but more-abstract decoratives and diagrams do not. The three design parameters for images are summarized in Table 1.

In Peircean terms, the design parameters of images are First Level in terms of form (unity) and reference (similarity), but the interpretation of images rises through the first level of feeling to also include the second level of active indication. That is, images may still be interpreted in terms of feeling, as all decoratives are, but images also have a second component of interpretation, which is the association with actual persons, places or things, that is, physical objects. Recall the shorthand code introduced in Chapter 1, Figure 10: we would refer to images as being the "one-one-two" type of visual, written as 1-1-2. The

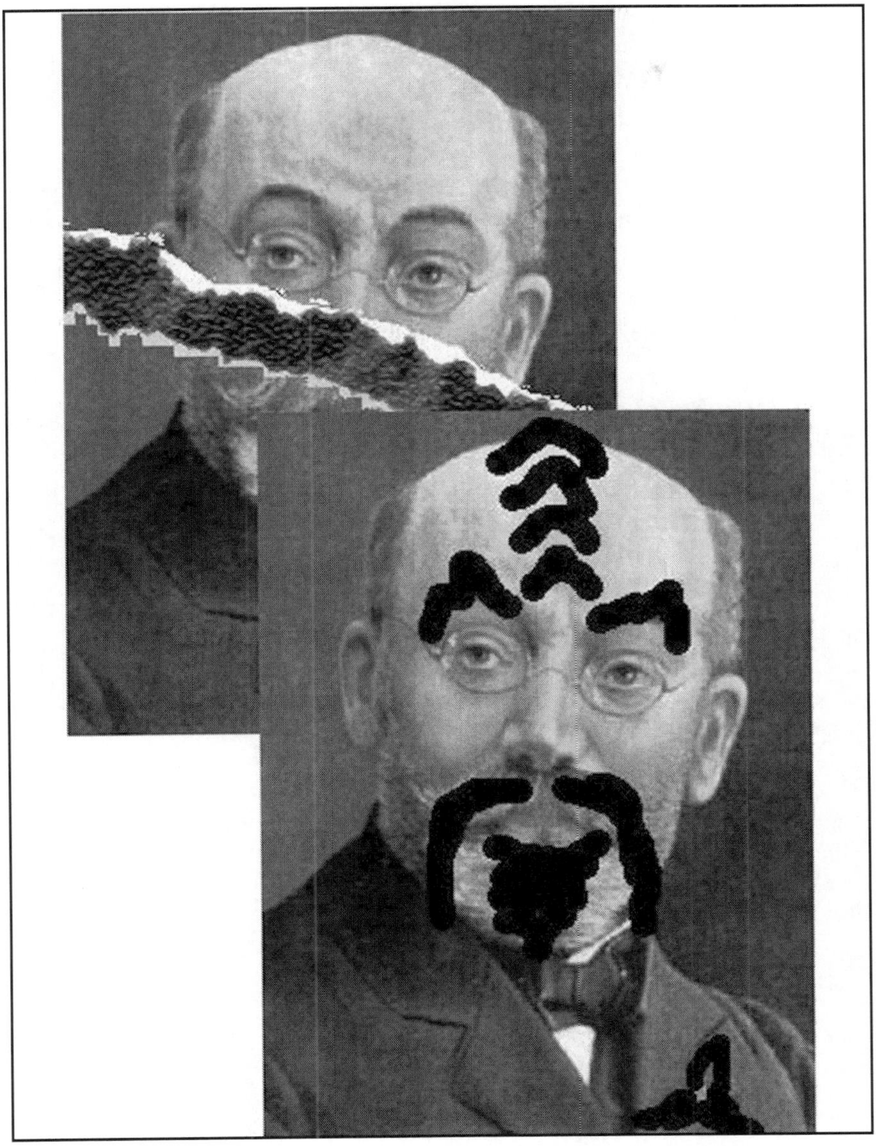

Figure 9. Vandalized images; disrupted unity.
commons.wikimedia.org/wiki/Image:Zamenhof_portrait.jpg

Broken unity disrupts basic image qualities and forces more complex interpretation, in this case, some kind of negative comment on the person portrayed.

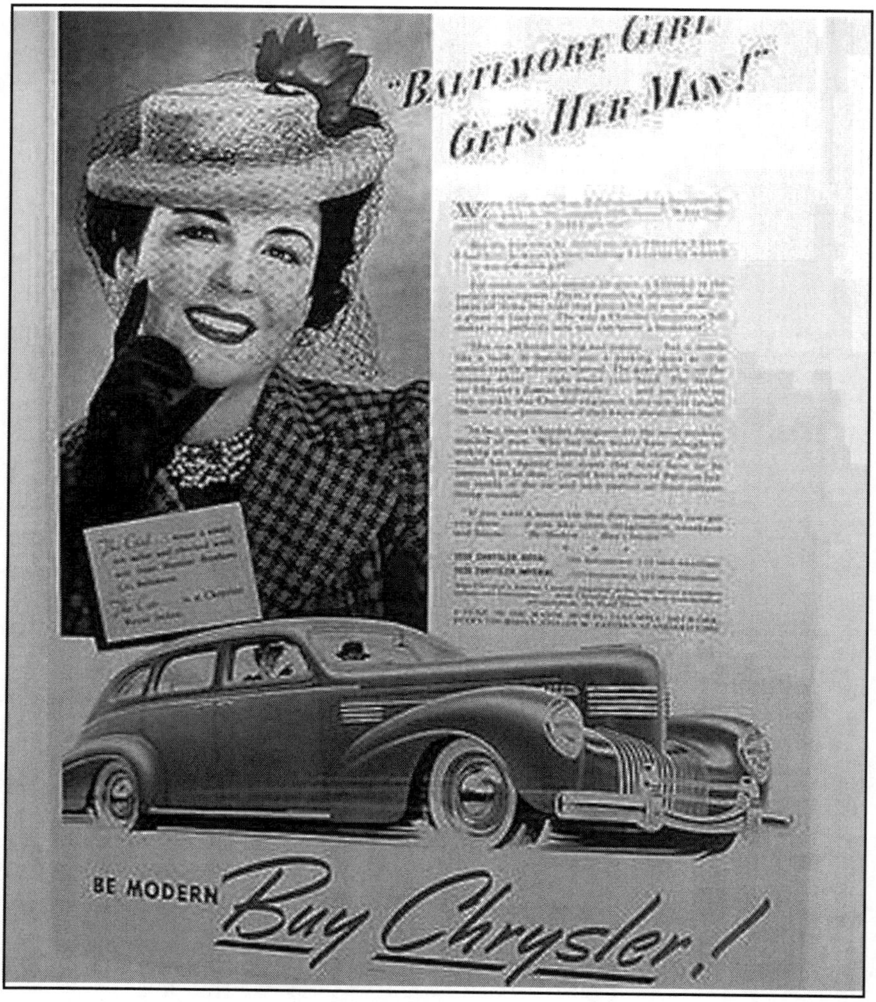

Figure 10. Image reference grounded in similarity to prior experience.

"Very politically improper by today's standards was this magazine ad for the 1939 Chrysler, which tells the story of how this 'Baltimore girl' won her husband by letting him drive her new '39 Chrysler. Throughout, she is referred to as 'girl,' while he is 'man.' The ad also reflects the prevailing magazine design of the day, using drawings instead of photos" (Wright, 2002).

Table 1. Key Design Parameters of Images; Shorthand Code = 1-1-2[a]

Basic Form	Referential Process	Interpretation Type
Qualities of unity	Qualities of similarity	Physical object connection
Form Level 1	Reference Level 1	Interpretation Level 1
1	**1**	**2**

[a]In Peircean terms, image design parameters are First Level formal unity and referential similarity, but Second Level for interpretation in terms of physical objects or qualities. Hence, the shorthand code, 1-1-2.

first "1" refers to Form, the second "1" to Referential Process, and the third number, in this case a "2," refers to Interpretation Type, as in Table 5. This shorthand code locates images relative to other visual types in Peirce's system of signs (Figure 11).

Recall that the Peircean system shows communication development, starting from the top left corner of the inverted pyramid, but progressing both downward and rightward. Therefore, the location of images in the 1-1-2 position (down and rightward from the pure decorative position in Figure 10) signifies the key properties of an image as a type of visual communication.

- Images may serve as decoratives, but to be most effective for that purpose, photographs typically require some adjustment "upward" to minimize connections to specific physical objects.
- Images mediate between pure feeling and physical action; they are felt as substitutes for persons, places, or things that physically affect viewers and so have great combined power to affect viewers both emotionally (as decoratives) and physically (as indicatives).
- Images can indicate as well as decorate, but unless they are supplemented with other forms (diagrammatic marks, captions, etc.), images lack any informative component and are therefore still prone to multiple, vague interpretations.

Because images are both powerful and readily available in the digital age, there is a marked tendency to overuse them or use them with little understanding of their likely effect on an audience. The problem lies not in the use of images *per se* but in their careless use or overuse. The solution lies in a more careful management of images—a conscious choosing and a conscious editing—shaping the image choices to a definite and an ethical purpose, conforming with their design properties of unity, similarity, and object connection. As Walsh points out,

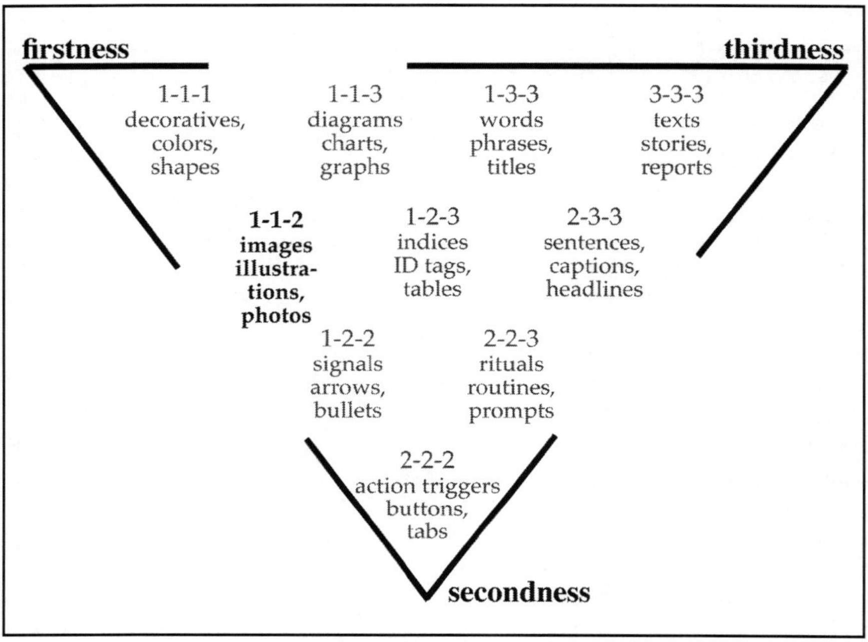

Figure 11. Images (boldface) located relative to other visual types (muted).

Peirce's 10-class system of visual types repeated from the Introduction and chapter 1 (Figures 2, 15), with coordinate numbers added, 1-1-1, 1-1-2, 1-2-2, 1-1-3, etc., shorthand for the level of form, reference process, and interpretation of each visual type. Image classification code is 1-1-2.

> The sheer complexity of a Baroque church relative to one from the Renaissance is best shown holistically in photographs. Likewise, a ceiling in the grotesque style needs to be experienced as a whole, not as a series of separate details. . . . Assuming an audience with a Christian background, the mere mention of themes such as "Magi" and "Madonna and Child" will evoke a sufficient theoretical framework to allow works based on these themes to be processed from photos alone. (1998, pp. 210–211)

Most generally, as Walsh suggests, images can be used to unify otherwise complex perceptions, and they can be effectively used to provide illustrative examples *after* orienting information has been delivered by other means, by diagrams or textual explanation.

IMAGE APPLICATIONS AND LIMITATIONS

Every single photograph or illustration that is pasted into a document should be evaluated to make sure its actual effects on viewers match the apparent purposes of the visual. It will often not be the case that audience need will align with the author's desire to showcase their graphical design expertise, to create a feeling of "wow" with the visual. In that case, the "wow" imagery will need to be deleted. Precisely because they do have power, images should be used cautiously rather than extravagantly.

In other words, images, with all their detail, need to be included in the computation of an effective "data/ink ratio" (Tufte, 2001), at least when the overall purpose of a visual is to inform. Even when the goal of the image is decorative or indicative, the details of the image still need to be carefully managed.

Photographs and Decorative Editing

If photographs are to function as decoratives, to represent feelings primarily, then the images have to be rather carefully manipulated. This is because a photograph, created through an automatic physical process, is a direct reflection of specific physical persons, places, or things. In Peircean terms, a photograph is an iconic image in terms of perception, but, in fact, it is created through an indexical process (CP 2.281). To function decoratively, an image's indexical connections to source objects have to be minimized, subdued, or otherwise muted by careful image selection, image processing, or image cropping.

Professional photographers routinely shoot dozens of images and from this large collection, they pick only one shot that best conveys the feeling they were looking to express. That one select shot will then be carefully cropped and perhaps altered digitally or chemically to enhance details that serve decorative purposes and to soften or eliminate details that don't conform to overall purpose. Such would be the case for the decorative images used in a Hawaiian travel brochure, for example. Conversely, the spontaneous photographs taken by actual tourists in Hawaii will normally be unsuitable as brochure-quality decoratives, as shown in Figure 12.

We see in Figure 12 that the default purpose of ordinary photos, such as the ones we take while on vacation, is indicative—to document specific events: birthdays, weddings, vacations, etc., and focus attention on specific people or things in specific places. Photographic images, indicative in their basic attributes, can be used only for primarily decorative purposes if there is careful framing and detail management. Most of the original image has to be cropped out in the decorative frame.

In the digital age, photographs and clip art illustrations are simple to acquire and paste into a text or slide presentation, using "drag and drop" or "insert" commands typical of Microsoft Word and other common word processing programs. These allow users to quickly and easily insert images into a document.

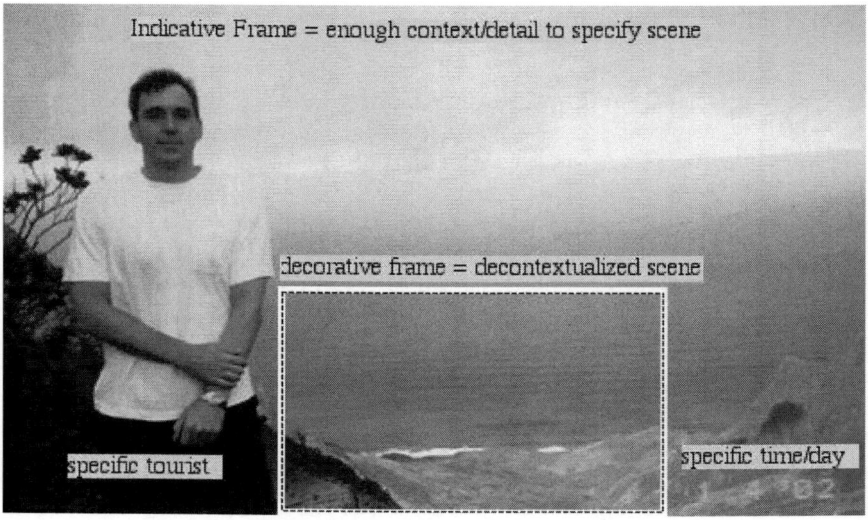

Figure 12. Indicative photograph manipulated for decorative effect.

Photographic images inherently reflect and indicate specific persons, places, and things. Photos must therefore be carefully manipulated and decontextualized for proper decorative effects.

That simplicity itself tends to promote carelessness, a failure to consider whether the graphic truly is appropriate for the overall purposes of the text.

On one hand, the author might intend an image to help an audience identify a particular object, but the clip art image from a standardized set may lack sufficient detail for indicative purposes. On the other hand, an author may intend an image purely for decorative purposes, but the clip art image may instead include details that overspecify the image (as in Figure 12) and interfere with the decorative purpose.

Decorative Effects in Images

If effectively edited to serve as decoratives, images can obey roughly the same form-feeling association principles as pure decoratives, assuming that the image has a dominant form and color scheme. Table 2 (recalling the discussion in the Introduction and chapter 1) outlines typical generic associations between dominant forms/colors of an image and particular ranges of feeling. As always, these effects can be overridden by strong personal or cultural associations with image content.

Table 2. Decorative Form/Feeling Associations in Images in the Absence of Specific Cultural Associations[a]

Basic Form of images	**Referential Process** Peirce-Category Similarity	**Interpretation Type** Range of Feeling
VARIETY and/or orange-yellow-green	Firstness: freedom, possibility, opportunity	AMUSEMENT
SHARP CONTRAST and/or orange-red-purple	Secondness: collision, action, danger	AGITATION
PATTERN and/or green-blue-purple	Thirdness: order, understanding, principle	FOCUS

[a]Form-Feeling associations for images obey the same principles as decoratives generally.

For example, an image of a desert scene with a variety of organic forms and textures—sand, rocks, and cliffs—in dominantly yellow or yellow-orange tones can readily connote freedom, pleasure, a sense of opportunity, and so on. These interpretations conform in turn with American cultural associations with the romance and freedom of the Western frontier (Figure 13, top).

Generally speaking, however, we can readily turn a desert scene away from these pleasurable feelings and push them toward the sense of the desert as a place of desolation or danger (another set of cultural associations) simply by sharpening the contrast levels of the image, muting the variety, and (if color is available) shifting color into the orange-red range (Figure 13).

Both the form and color scheme of a decorative image can align with a single Peircean category. Also, mixed effects can be created, parallel to the discussion of pure decoratives in chapter 1. For example, a typical photo of Earth from orbit is dominated by blue-range color, but ocean, clouds, and continents create considerable formal variety. The net effect of this pleasure-resolution combination is (usually) a general feeling of peace, tranquility, a oneness with nature, etc., which again aligns with cultural associations with space-related images.

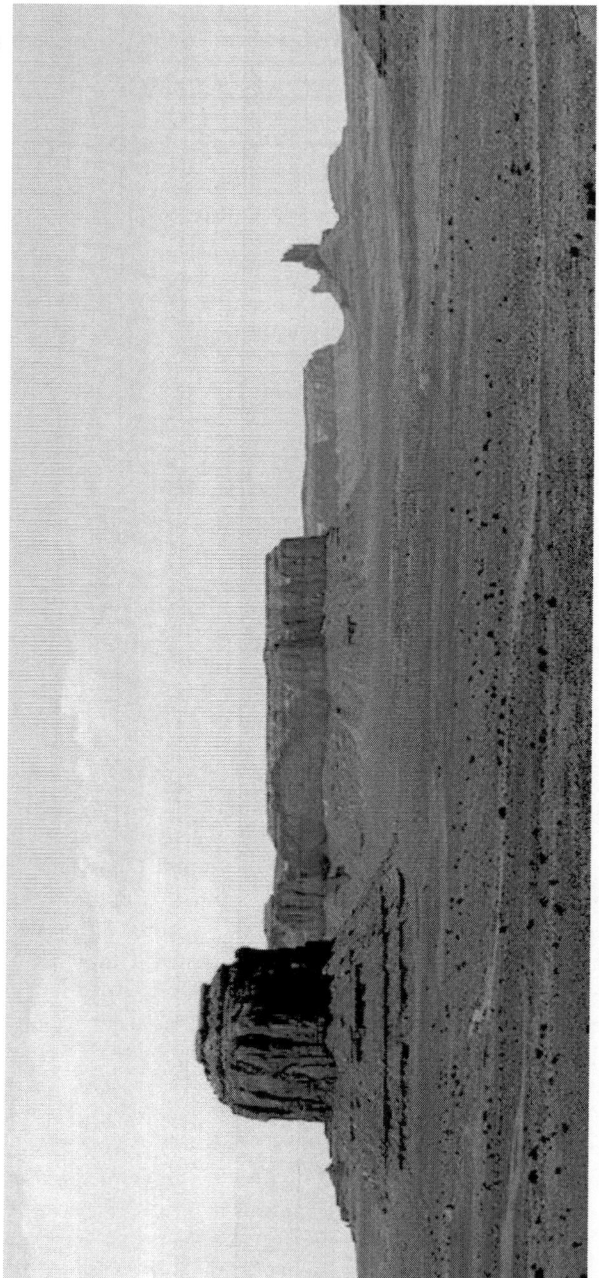

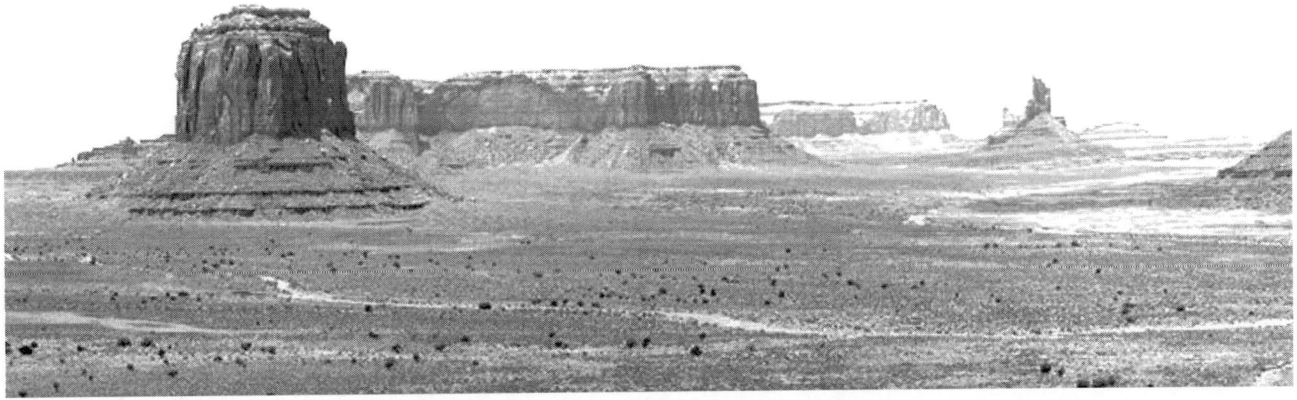

Figure 13. Image variety, contrast, and color effects.
commons.wikimedia.org/wiki/Image:Artist_Point_Monument_Valley_USA.jpg GNU license. Photo by Tobi 87.

Decorative feelings in an image can be modified by adjusting contrast or color (not directly shown here).

Action Mediating Effects in Images

Because images have both decorative and indicative properties, they serve as essential intermediates between feeling and action. Consider, for example, our reaction to any raw perception, say, a light, tickling feeling on the back of one's neck. Is this a good feeling? Bad feeling? Neutral feeling?

Whatever image instantly comes to mind mediates how we will react; the actions that follow from the feeling depend on what object-image forms in the mind in conjunction with the pure feeling, whether you imagine the light touch of a lover or a loose thread from your collar or, perhaps, the image of a free-roaming spider. Regardless of the real cause, the mental image determines whether you will sigh with pleasure or jump in terror in reaction to one and the same feeling, which is neither good nor bad until you've tied it to an image.

To take another example, suppose you own stock in a company. You've just found out through insider information that the stock is about to fall. Your basic feeling is concern about losing all that capital. Whether you act on that feeling and sell the stock is mediated by whether or not you have the clear image in mind of going to jail for insider trading and how that image weighs against a competing image of cashing in the stock and, perhaps, sailing to Bermuda.

People can be induced to accept change (or resist it)—to act or not to act—when presented with images that encapsulate their feelings and connect those feelings to concrete, physical interpretations, interpretations that typically lead to decisions to achieve or avoid the outcome reflected in the chosen image. From this core attribute—the mediation between feeling and action—emerges the principled use (Figure 14) or unprincipled use (Figure 15) of images in advertisement and political persuasion.

Because images mediate between feeling and action, they are extraordinarily useful as motivational tools, but they are, for that same reason, also extraordinarily prone to ethical abuse. People can be manipulated by images that seem indicative of actual outcomes, desirable or frightening, but are in fact disconnected from reality. For example, it is a common practice in fashion magazines to elongate a raw photograph slightly, such that a female model appears somewhat taller and thinner than she actually is. This and several other techniques are used to create a decorative iconography of feminine beauty that is far out of step with the way that even the fashion models themselves normally look (Reaves, Hitchon, Park, & Yun, 2004).

The actions provoked by this manipulated imagery are favorable to the fashion industry but socially corrosive in most other ways. Some women purchase figure-slimming clothing, diet aids, and exercise equipment in endless cycles of wasted time, money, and distress simply because the sought-after images seem to exist but in fact are unreachable.

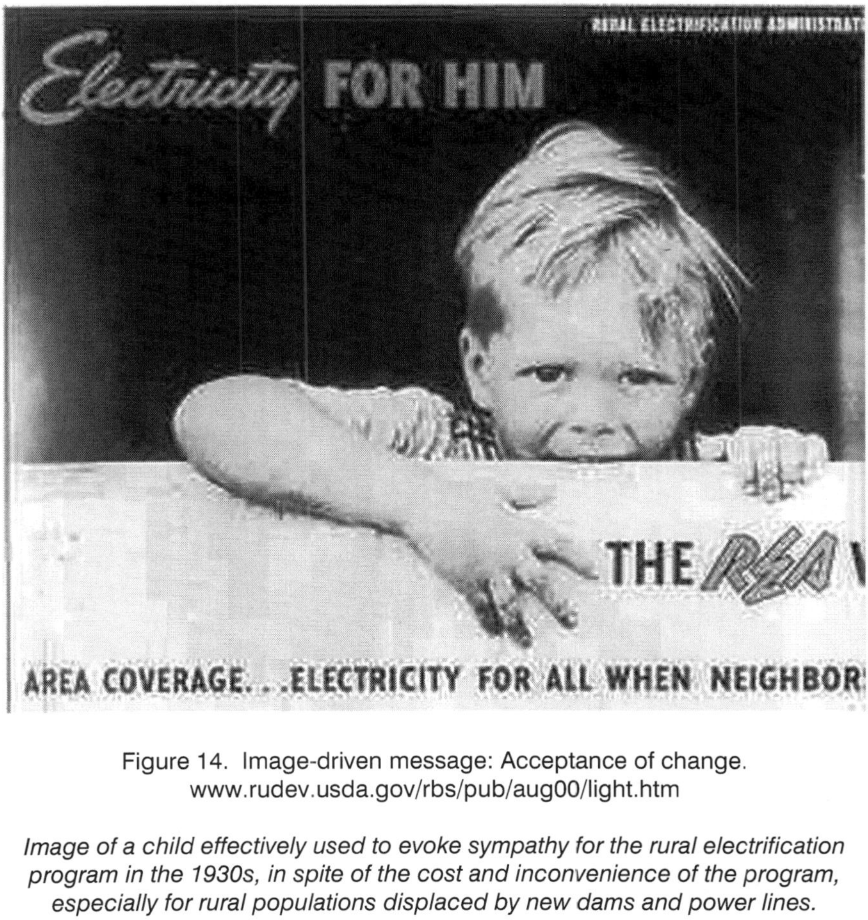

Figure 14. Image-driven message: Acceptance of change.
www.rudev.usda.gov/rbs/pub/aug00/light.htm

Image of a child effectively used to evoke sympathy for the rural electrification program in the 1930s, in spite of the cost and inconvenience of the program, especially for rural populations displaced by new dams and power lines.

LIMITATIONS OF AND ALTERNATIVES TO IMAGES

Realistic illustrations and photographs—images in the narrow, technical sense—acquire meaning by virtue of physical resemblance to objects. Non-imagistic but nevertheless iconic forms (e.g., abstract shapes, textures, cartoons, and diagrams) likewise acquire meaning through resemblance but not necessarily resemblance to physical objects. Abstract decorative forms are referenced by similarity to feelings evoked, and diagrams are referenced by similarity to conceptual relationships not ordinarily perceived directly (like borders on a map).

To review the full range of icon choices available, recall from chapter 1 that pure decoratives resemble only feeling-features abstracted away from objects

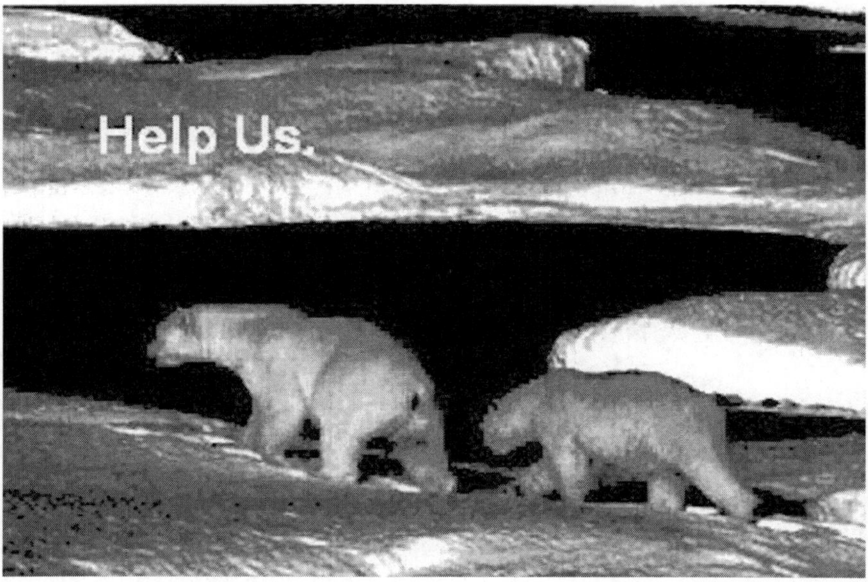

Figure 15. Exemplary image-driven political message.

Overtly manipulated image (extreme red shading not shown) used to provoke political and personal choices to reduce global warming.

(lines, shapes, textures, etc.) As will be discussed in chapter 3, cartoons and other diagrams resemble the ideas associated with objects rather than the objects themselves. The cartoon *Felix the Cat,* for instance, resembles no real cat and is therefore not an image in the strict sense. The cartoon cat instead resembles and represents the idea of a thinking, talking, trickster figure in a cat-like guise. In sum, each of these three icon subtypes (decoratives, images, and diagrams) resembles a different aspect of reality, with clearly overlapping but nevertheless distinct ranges of suitable uses. The full range of possible icons is represented in Figure 15, an elaboration of Figure 1 above, with alternative examples.

The threefold distinction between pure forms, images, and diagrams is also discussed at length by Scott McCloud in *Understanding Comics* (1994). McCloud points out that cartoon or comic figures are one lateral step away from pure feeling-evoking forms, as shown in Figure 15, top. The tiger-stripe border or the stylized cats eyes represent form-features that are abstracted away from any particular source but that do evoke feelings associated with cats,

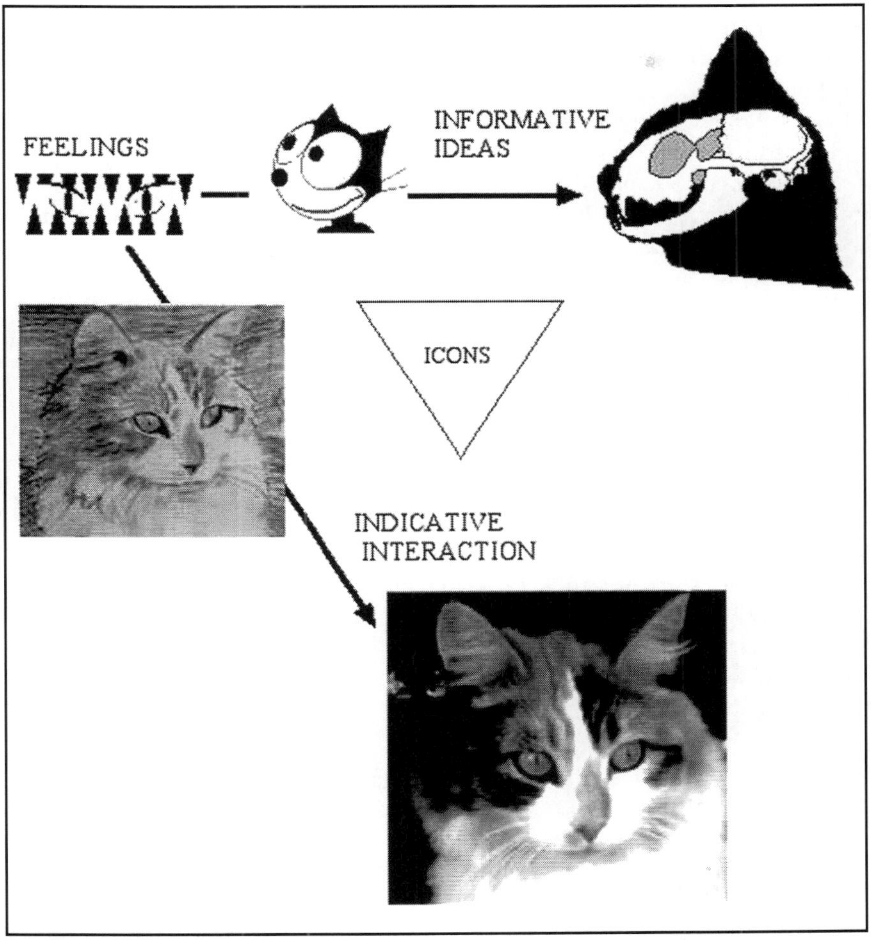

Figure 16. The range of icon types. Besides images, there are abstract decoratives and diagrams.

Besides indicative images, there are feeling-provoking decorative forms and informative diagrams. Cartoons are intermediate between diagrams and decorative forms; illustrations are intermediate between decorative forms and indicative photographic images.

but even more generally, danger or mystery or intrigue. Cartoons move away from abstracted forms but in the direction of another kind of abstraction: the ideational relationships most purely represented by informative language. Along that continuum between pure forms and language, we find cartoons and, a little farther along, we find informative diagrams.

Realistic illustrations and images move in an entirely different direction, down from the abstract levels of either form or relation, down toward a more concrete reference to specific objects. To move downward in the Peircean triangle, away from form/feeling abstraction, is to move toward the specific, the physical, and the concrete.

The cat illustration reflects one specific variety of cat (the long-haired calico), whereas the cat photograph reflects one unique cat (the second author's). Cat illustrations and cat photographs may still represent cats in general, or even feelings associated with cats, but to do this, the images must be carefully framed to eliminate contextual detail, as was discussed earlier. In Figure 16, the cat illustration is placed against a white background, and the cat photograph is carefully cropped to exclude the specific owner, who is holding this specific cat.

Ultimately, images are grounded in the physical world, and this grounding, however useful and captivating, also creates fundamental limitations. The physical world is full of infinite detail, not all of which can be relevant to a finite communicative purpose. The overall shape of the proverbial forest is not seen due to all the distracting details of individual trees. Most messages are made clear only when distracting details are discarded. Images, by their nature as physical reflections, carry with them the burden of physical detail.

CONCLUSION: TOWARD AN IMAGE ETHIC

Our particular concerns in this book are the clarity and suitability of visual design strategies relative to specific goals. Our particular concerns in this chapter have been the clarity and suitability of images chosen for various kinds of visuals.

As discussed in the Introduction, ethical action is here defined as action directed toward sustainable, attainable goals: goals acceptable to both the actor and all those who are acted upon. As a subdomain of visual communication, image-driven communication is an even more specific domain where this general model of ethics applies: the one presenting an image acts upon viewers of the image, to accomplish whatever goal that all are collectively pursuing, assuming that the goal is attainable by the means chosen and sustainable over the long term.

- This definition of ethical action covers the obvious requirement that images be accurate, in the specialized sense that they do not mislead viewers about the real facts.

It is the nature of images that they do not in themselves assert propositions, so they cannot of themselves be directly true or false. It is also the nature of images that they can only selectively reflect reality, whatever specific aspects of an object that an illustrator chooses to draw, or whatever the photographer chooses to frame and focus on. As we have seen, photographic images have to be edited and usually processed to achieve specific decorative effects.

Thus, no image directly asserts truth, and no image can directly lie. The ethical challenge lies in presenting images that do not mislead. We have consequently discussed instances where images can be misleading, especially when merged with otherwise accurate information but merged in distracting ways (Figure 7). Whether or not the presenter of an image is being deliberately deceptive, an image that is generally misleading in its effects fails to ethically meet the goal of the community collectively, to discover the truth.

The Peircean definition of ethical action covers what should be the equally obvious requirement that a visual not harm any of its viewers or harm the wider community acted upon by what the viewers of the visual do. This definition of ethical action also covers less obvious requirements:

- That visuals be as clear as possible in their information; as clear as possible in what they indicate, and in what feelings they evoke; otherwise, the goal of viewers to understand what they're looking at is frustrated.
- That the strategies used to construct visuals be suited to actually attaining their apparent purposes; that images, for example, not be used by themselves to directly convey conceptual information; otherwise, informative goals of both the visual creators and viewers, for that particular visual, are unsustainable.

In general terms, it is less than ethical to use images in uncontrolled, inconsistent ways, such that feelings and object associations evoked by the image are out of line with the overall purpose of the visual (as in Figure 5). When designers highlight, with the deployment of images, their personal prowess with image placement and manipulation for an audience that is primarily seeking clear information, there is a mismatch of goals and an ethical breach.

Figure 17 exemplifies several of the difficulties in the proper and ethical selection of images, as previously discussed. Again, we find images interfering with easy navigation of the U.S. Library of Congress Web site.

As shown in Figure 17, the designer has taken considerable trouble to create a flash-animation-driven menu. Imagery taken from several different library resources is combined in a single, densely colored collage. As each region of the collage is "rolled over" with a mouse pointer, a different resource is highlighted. In this instance, the Library's image collection is highlighted. Unfortunately, this highlighted portion clashes with the background collage. Moreover, the

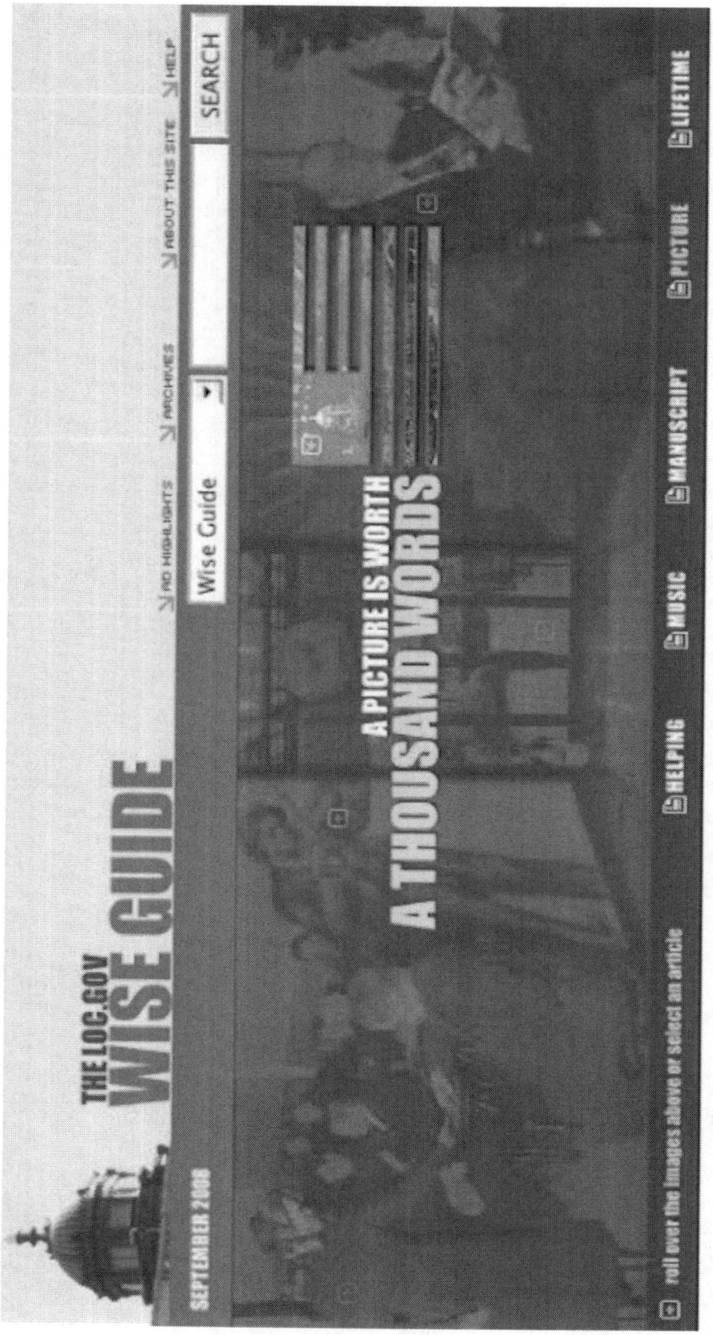

Figure 17. Image-cluttered navigational menu.
http://www.loc.gov/wiseguide/sept08/index-flash.html

A variety of design problems are combined in this navigational menu: color-feeling associations are problematic, formal unity is disrupted when specific elements are "moused over," and images do not transparently correspond with distinct library resources.

dense red of the collage is somewhat more agitating in feel than one might expect from a representation of library holdings.

Though technically impressive, the navigational purpose of the visual has been badly sidetracked. To some degree, we can excuse image-choice problems when they are not deliberate, but only to the same degree that we excuse factual inaccuracy from someone who does not intend to deceive. In any case, an ethical responsibility rests with the visual designer for the overall effects of the visual, in which images may play either a positive or an obstructive role. Conversely, positive ethical value is manifest whenever a visual designer has obviously taken care to match images to a specific emotional or indicative purpose (recalling, for example, Figure 14). Comparable, positive considerations are again illustrated in proposed revisions to the Library of Congress page, as shown in Figure 18.

Further considerations in the construction of such menus will be discussed in chapters 4 and 5. In the next chapter, we will discuss the general importance of extracting image-like details from diagrams so that conceptual relationships, the essence of information, can be made clearer.

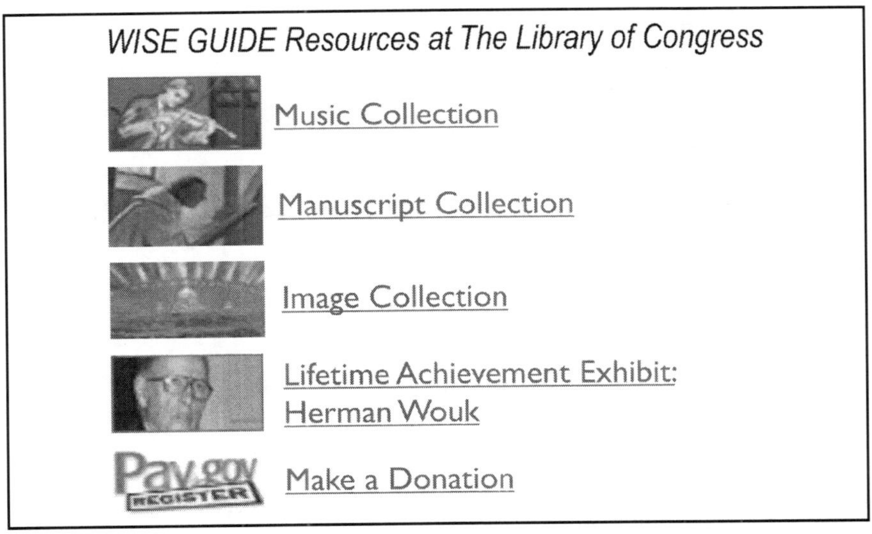

Figure 18. Image-removed informative graphic (more effective).

In the proposed revisions, images are minimized and separated so that they serve less of a decorative purpose and more of an indicative purpose, as is appropriate to a navigational menu. When images are used in this way, it is important to make sure that the image actually used will transparently correspond to adjacent words indicated.

CHAPTER 3

Diagrams

Diagrams, charts, and graphs have strong decorative (i.e., aesthetic) considerations but also naturally have inherent informative properties. In the Peircean analysis, true informative purpose can be attained only under specific conditions where the essential visual properties of information are enhanced: (a) redundant and extraneous details stripped away, (b) contrasts between critical details emphasized, and (c) generalizations represented in the graphic validated by collateral textual support. If these conditions aren't met, then the diagram/chart/graph will degrade into a mere image of information, a decorative-indicative substitute for information, most typically, an impressive looking graphic that few or none in the audience can understand. We will use the term *diagram* to cover all the usual senses of that term (Dictionary.com):

1. a figure, usually consisting of a line drawing, made to accompany and illustrate a geometrical theorem, mathematical demonstration, etc.
2. a drawing or plan that outlines and explains the parts, operation, etc., of something: e.g., a diagram of an engine.
3. a chart, plan, or scheme.

But, in addition to these usual kinds of diagrams, we will use the term *diagram* to refer to other kinds of visuals that correspond with Peirce's type V visuals (shorthand code 1-1-3).

In other words, a diagram is any visual that represents by similarity as an icon but that is interpreted in terms of conceptual, conventionalized similarities (i.e., as *ideas*) rather than in terms of feeling (as decoratives are interpreted) or as reflecting specific physical object perception (as images are interpreted).

> A diagram . . . is any general law or type, in so far as it requires each instance of it to embody a definite quality which renders it fit to call up in the mind the *idea* of a like object. (CP 2.258, emphasis added)

> Diagrams and diagrammatoidal figures are intended to be applied to the better understanding of states of things, whether experienced, or read of, or imagined. Such a figure cannot, however, show what it is to which it is intended to be applied. (CP 3.419)

By this definition, a map is also a diagram, a cartoon is a (decorative-leaning) diagram, an algebraic formula is a (language-leaning) diagram, a bar or line graph corresponds with what Peirce calls a "logical diagram" (CP 3.619), and so forth. In each case, diagrams are forms that are like what they represent, as ideas (for example, hypothetical piles of all the steel produced by each country in a given year), but at the same time, diagrammatic forms do not look like physical objects to which the diagram "is intended to be applied."

The term *diagram*, then, is here used to designate all visuals constructed to resemble objects or ideas more in their conceptualized, contrastive, part-to-part relationships than in their ordinary physical appearance. Figure 1 provides further examples of this decorative-image-diagram distinction.

Readers will recall that diagrams were referenced briefly in chapter 2 in order to clarify key attributes that differentiate this visual type from the type referenced by the term *image* (recalling Figures 1 and 16).

Primary attributes of conceptualized, contrastive, part-to-part resemblance are what define diagrams, in general, differentiated from decorative forms that lack conceptual content but emphasize feeling and minimalize contrasts (Figure 1, top left). Similarly, images (Figure 1, bottom center) reflect the feelings associated with objects, while the sheer level of detail in images tends to blur or obscure relationships between distinct parts of objects. Cartoons, maps, and diagrams emphasize and indeed exaggerate contrasts abstracted from knowledge (rather than physical appearance), and they further emphasize these contrasts by discarding physical details that do not support those contrasts.

A typical cutaway diagram showing the internal workings of, say, a flower, does not look like any actual flower that could successfully go on living and growing, being thus cut in half and having key reproductive parts exposed (Figure 2).

Whereas an image reflects more or less exactly what we would see if we looked at something in reality, a diagram shows us what we ordinarily would *not* see just by looking at something in reality. A few exceptions to this rule occur when a diagram has been constructed out of physical materials—like a foam-rubber suit being shaped to look like a cartoon character, for example, or an actual aircraft's fuselage and engine housing being cut away, some parts removed and others painted bright colors to expose interior details, or a skeleton carefully mounted and displayed in a museum. In such cases, any photographic image of such deliberately crafted materials may have a diagrammatic function. Histology slides commonly used in medical research are another example of materially constructed diagrams (Figure 3).

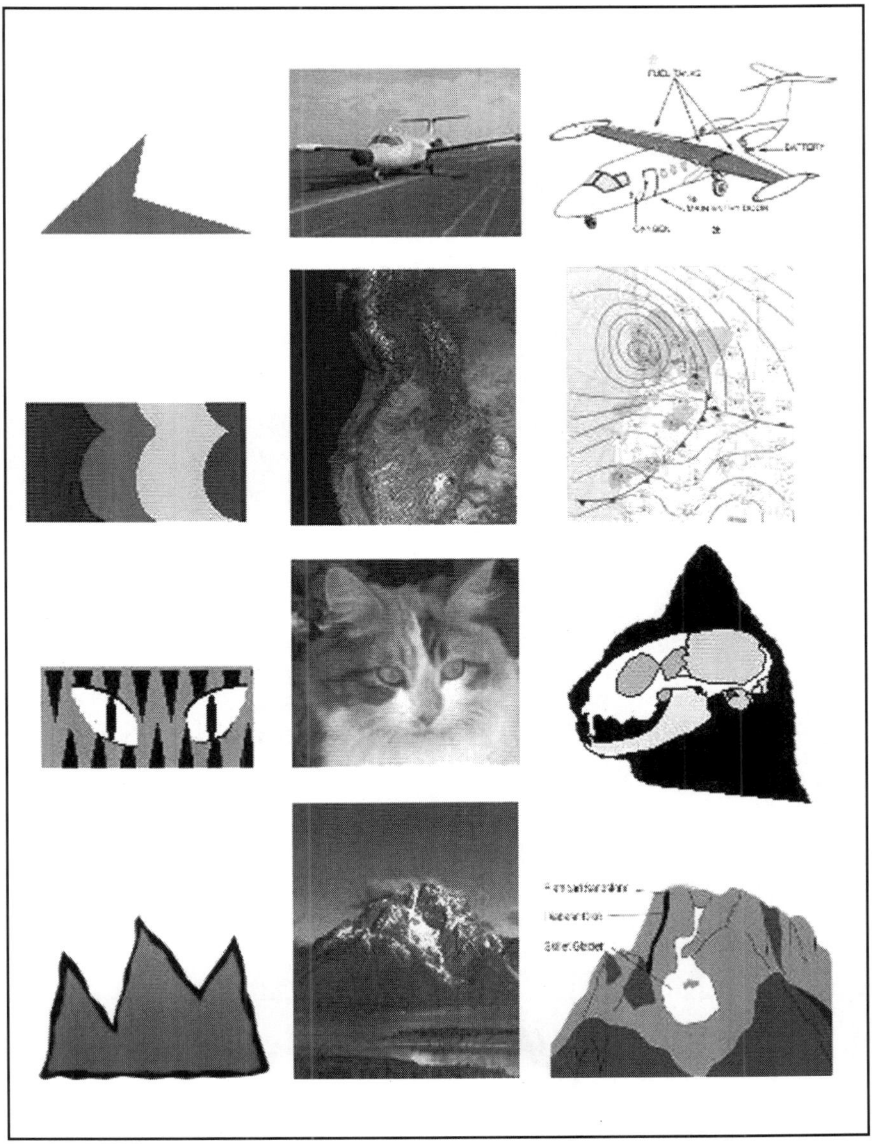

Figure 1. Pure decoratives vs. images vs. diagrams.

Decorative forms, images, and diagrams are all interpreted by similarity, but a different kind of similarity defines each type: similarity to feeling defining decoratives, similarity to objects defining images, and similarity to conceptual relationships, i.e., ideas, defining diagrams.

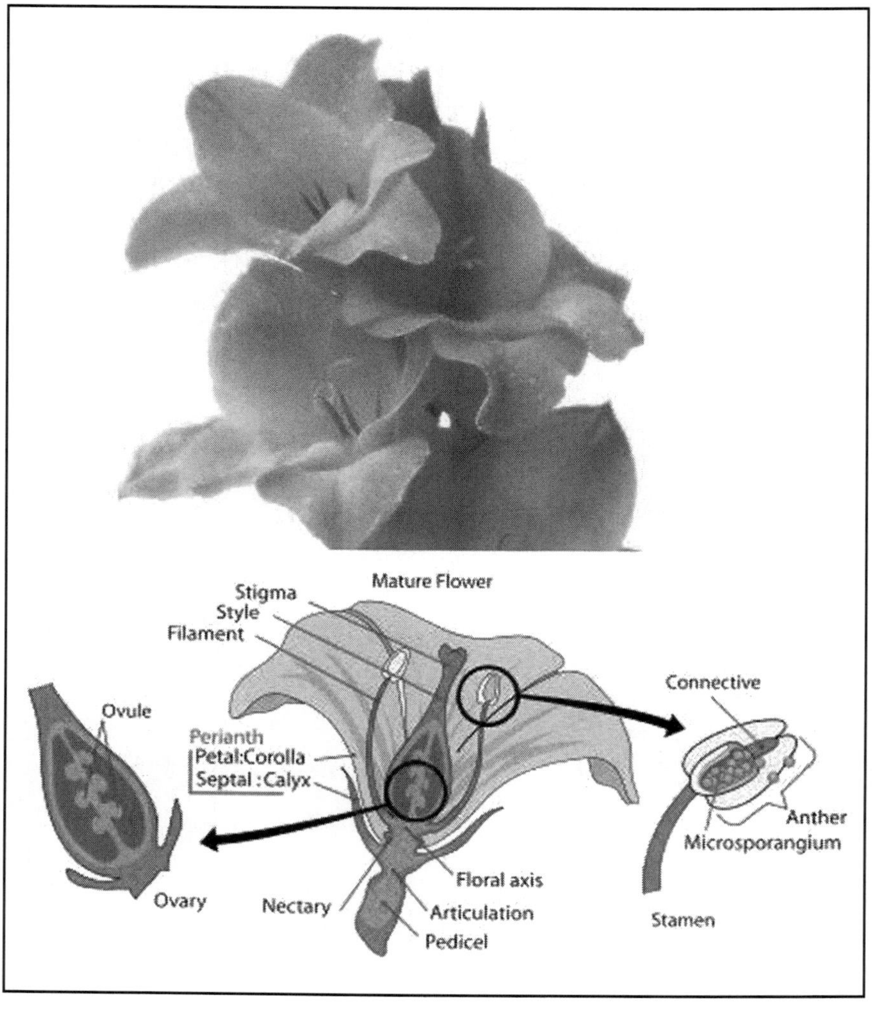

Figure 2. Image vs. diagram of the same kind of object.

Whereas an image reflects more or less exactly what we would see if we looked at something in reality, a diagram shows us what we ordinarily would not see just by looking at something in reality.

In this chapter, we will discuss diagrams, defined not by their diverse methods of construction but rather by their focus on relevant detail (and exclusion of irrelevant detail), by their emphasis on contrast, and by their general informative function, as opposed to the decorative and indicative functions of images, which generally lack stark contrasts and in which relevant and irrelevant details intermingle.

DIAGRAM DESIGN—PEIRCEAN PARAMETERS

Recall that each visual type has three key parameters, based on the Peircean categories:

- its basic form—whether the visual is unified or has distinct phases
- its referential process—how the visual acquires its meaning
- its interpretation type—what kind of meaning the visual form typically acquires.

Basic Form

Successful diagrams require significant but subtle decorative touches. These visuals should attract the eye but at the same time put viewers at ease, such that they feel willing and able to inspect the visual and put forth the effort to extract information from it. Diagrams, like images and pure decoratives, must convey a sense of visual unity or balance. In other words, all three of these visual types share the level-1 form parameter and therefore obey visual gestalt principles. Beyond these basic requirements for balance, symmetry, and continuity, however, diagrams generally work best if they do not contain many overtly decorative or image-like elements, especially colors or shading touches, that distract from their informative core.

Diagrams with image elements superimposed are the most frequent object of criticism in Edward Tufte's classic *The Visual Display of Quantitative Information* (2001). Such visuals are referred to as *chartjunk.* Note that image elements here (Figure 4) include 3-D effects, as if the graphic figures had physical depth.

In sum, the formal unity of diagrammatic form is defined in reference to a unified block of information. Unnecessary imagistic or decorative features, not directly supporting that information, will disrupt the genuine unity of the diagram, as illustrated in the chartjunk example.

Referential Process

Like decoratives and images, a diagram gets referred to its interpretation through qualities of similarity. We learn to "read" decorative feeling and images reflecting physical objects simply by virtue of living in a world of feelings and physical objects. The experiences needed to understand diagrams are much

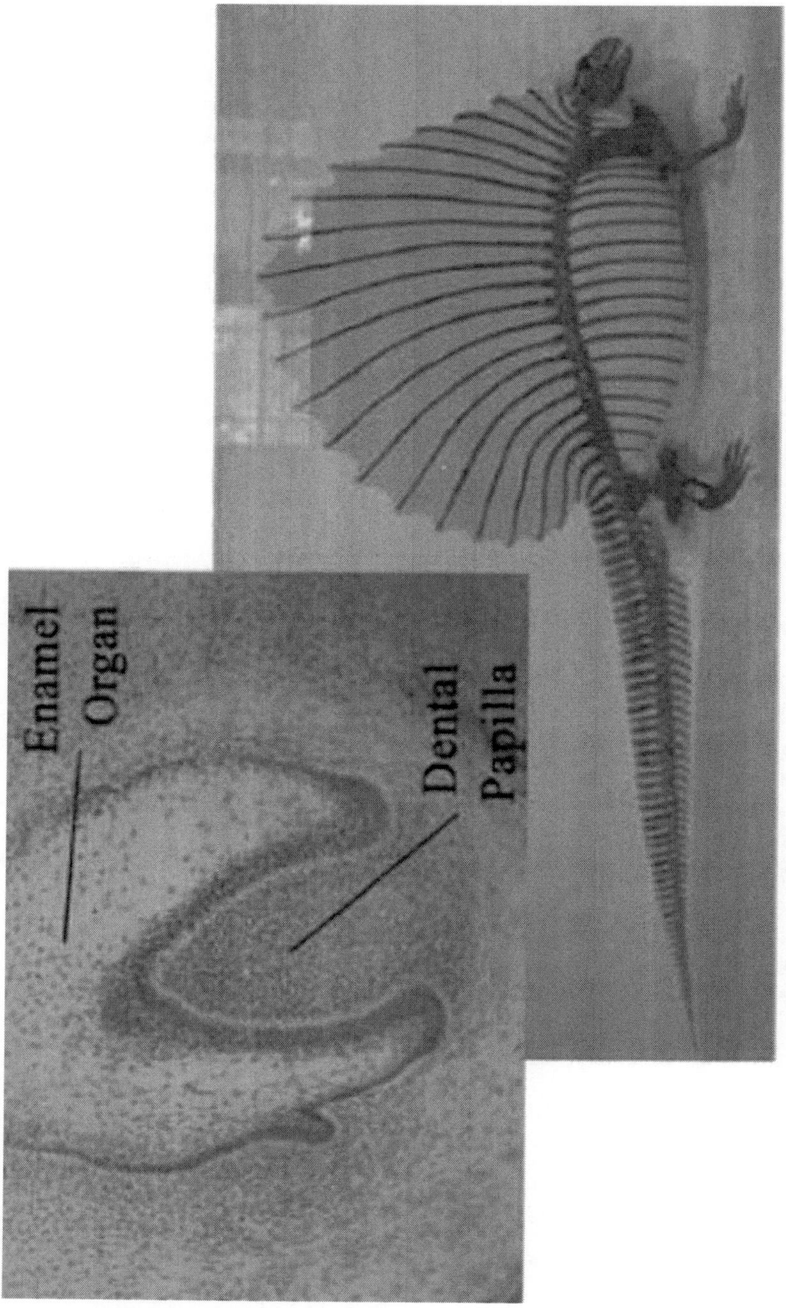
Enamel
Organ
Dental
Papilla

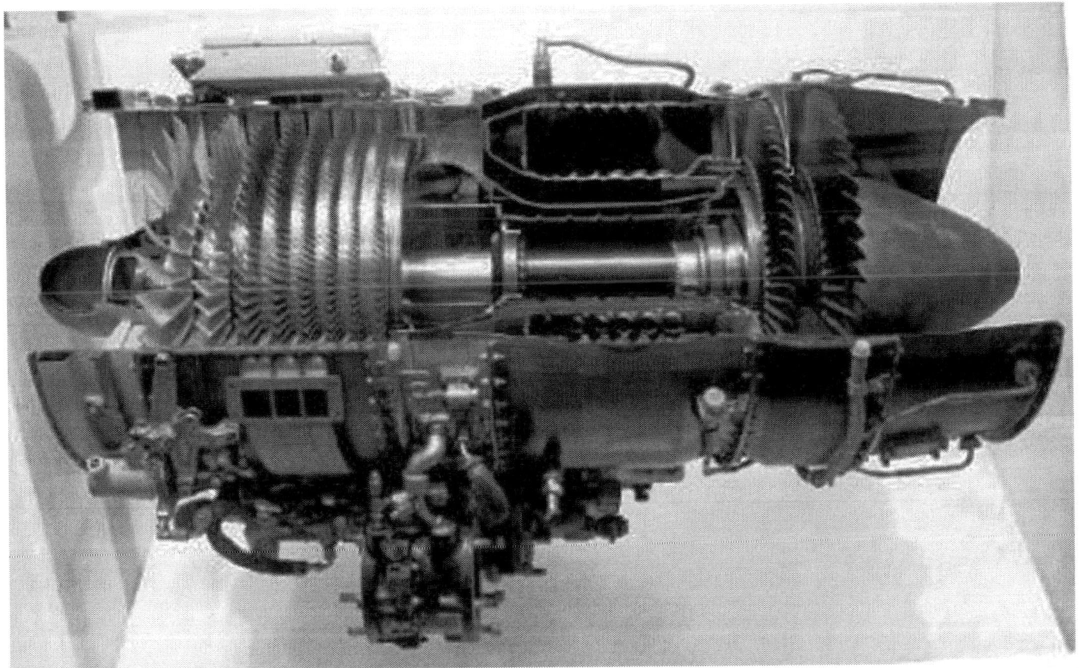

Figure 3. Photos of materials constructed with diagrammatic function.

Diagrams can be made out of physical materials but still do not "look like" how these objects would naturally appear in ordinary circumstances.

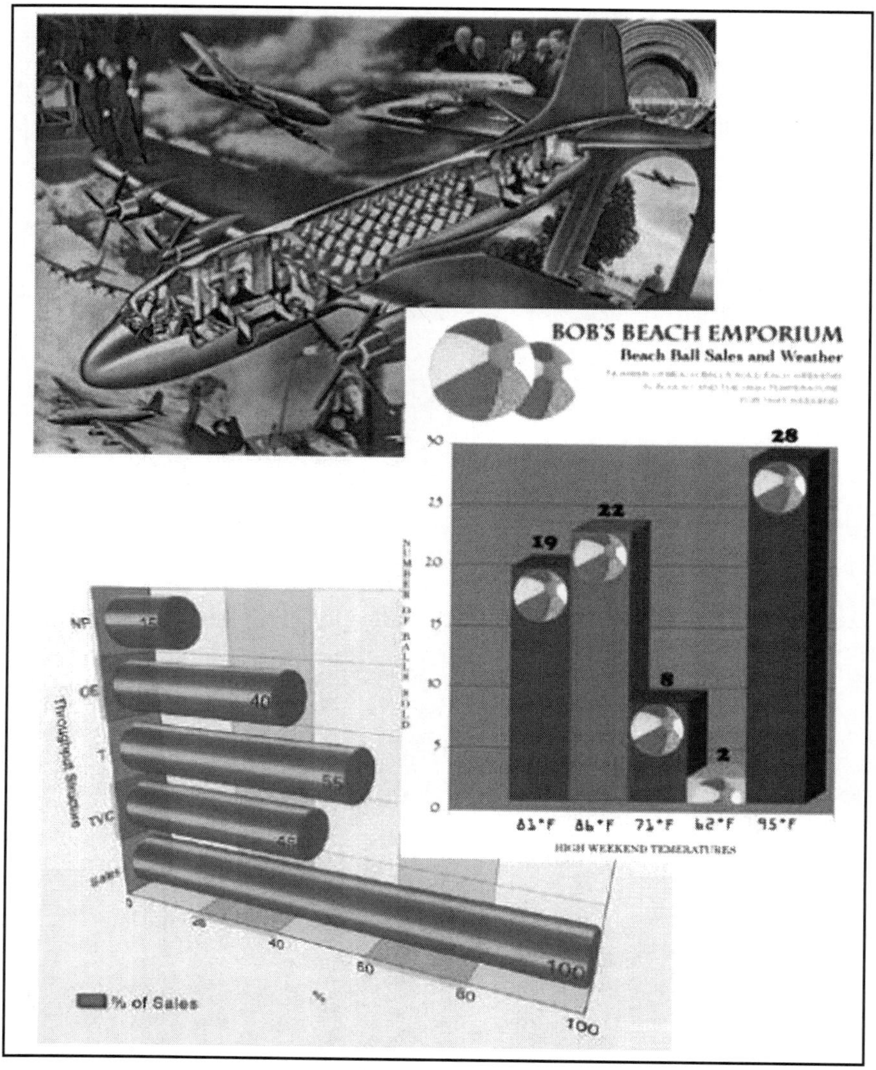

Figure 4. Some classic chartjunk examples, a violation of information unity.

In chartjunk, image elements intrude on diagrammatic forms and purposes.

less common in the natural world, however. We therefore generally require specialized instruction in order to read diagrams.

In other words, just as we understand images by association with our constant experience with physical perceptions, we can only learn to read diagrams based on a number of initiating experiences with similar conceptual comparisons. Not surprisingly, diagrams in the field of design are often called *informative* or *information graphics*, most likely because the primary goal of diagrams is to inform.

For example, this is why most children have to be shown how to read a clock or a map, initiated into the conventions associating hands and numbers with increments of time, or the conventions associating lines and colors with landmarks and regional boundaries. Similarly, it has become part of basic Western education to learn how to read a graph, or table, and part of more specialized education to learn to read circuit diagrams, chemical diagrams, and so on.

The message of a diagram, more often than not, depends to a large degree on viewers' prior initiation into certain "codes," specific strategies of information display, specific analogies between conceptual relationships like relative quantity and form relationships like relative height of bars in a chart. Even so, this prerequisite initiation into diagram codes is not different in principle from the prerequisite associative experiences in the physical world that determine how viewers will interpret images or decorative forms. What varies between decoratives, images, and diagrams is exactly what kind of experience is being associated with each visual type: feeling to decoratives, physical objects to images, and conceptual relationships to diagrammatic forms as shown in Figure 5.

Diagram design choices, like decorative and image design choices, can therefore misfire if the author misjudges the prior education of viewers, education that determines the associations that viewers are able to make with the diagrammatic icons presented.

The definitive property of icons, whether decoratives, images, or diagrams, is that interpretation emerges from perceived similarity, based on prior experience of viewers, but this prior experience, and thus the meaning of any icon, can certainly vary. The challenge for any visual communication designer is to anticipate possible variations in the interpretation of a visual and to provide enough supporting information and context to go with the visual, such that the intended interpretation corresponds with actual interpretation in as many cases as possible.

Interpretation Type

The interpretive goal of diagrams is an informative one: the assertion of propositional information. This is what differentiates diagrams from the other icon types, decoratives, and images, which are otherwise alike in terms of unity of

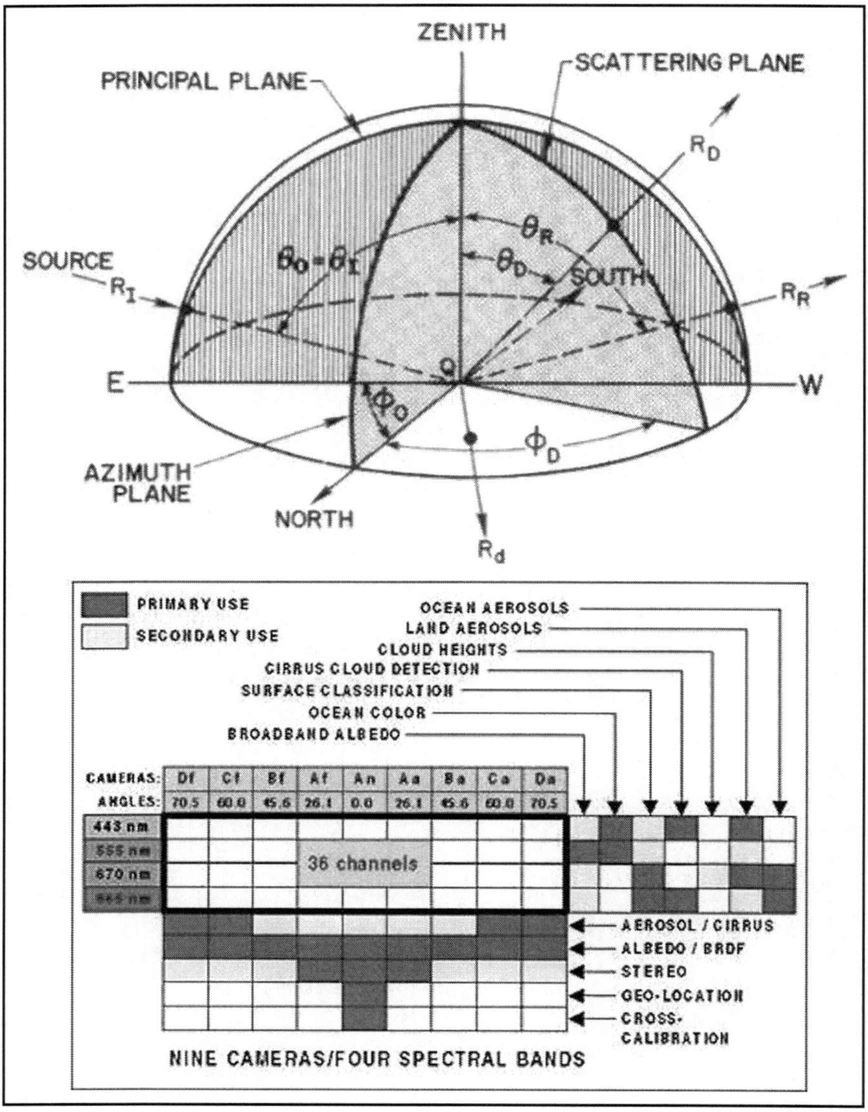

Figure 5. Well-formed but difficult-to-interpret diagrams.

Diagram reference, like image reference, is grounded in prior experience. For images, the prior experience is physical; for diagrams, the prior experience is the learning of conventional codes.

form and reference by similarity. The three design parameters for diagrams are summarized in Table 1.

Speaking in terms of the grammar of visuals, the third interpretive level implies the second and first levels. In other words, level-3 diagrams may still evoke feeling, as level-1 decoratives do, or indicate objects, as level-2 images do, but diagrams also have a third component of interpretation, which configures informative relationships.

Recall the shorthand code from chapter 1, Figure 10, and chapter 2, Figure 11. We would in the same mode refer to diagrams as being the "one-one-three" types of visual, written as 1-1-3. The first "1" refers to Form, the second "1" to Referential Process, and the third number, in this case a "3," refers to Interpretation Type, as in Table 1. This shorthand code locates diagrams relative to other visual types in Peirce's system of signs (Figure 6).

Recall that the Peircean system shows communication development, starting from the top left corner of the inverted pyramid but progressing both downward and rightward. Therefore, the location of diagrams in the 1-1-3 position (rightward from the pure decorative position and upward from the image position in Figure 6) signifies the key properties of a diagram as a type of visual communication.

- Diagrams are the first intermediate step between image-like perception and knowledge expressed in language. Diagrams act as a bridge to convert perception into verbal information and vice versa: verbal information into perception.
- Diagrams not properly configured as language-perception bridges can easily "degenerate" into merely decorative visuals or into images of information that viewers can perceive but not process or interpret as information.

Table 1. Key Design Parameters of Diagrams; Shorthand Code = 1-1-3[a]

Basic Form	**Referential Process**	**Interpretation Type**
Qualities of unity	Qualities of similarity	Informative relationships
Form Level 1	Reference Level 1	Interpretation Level 3
1	**1**	**3**

[a]In Peircean terms, the design parameters of diagrams are First Level in terms of form (unity) and reference (similarity), but the interpretation of diagrams rises through the first level of feeling and the second level of object reference also to include the third level of informative relationships.

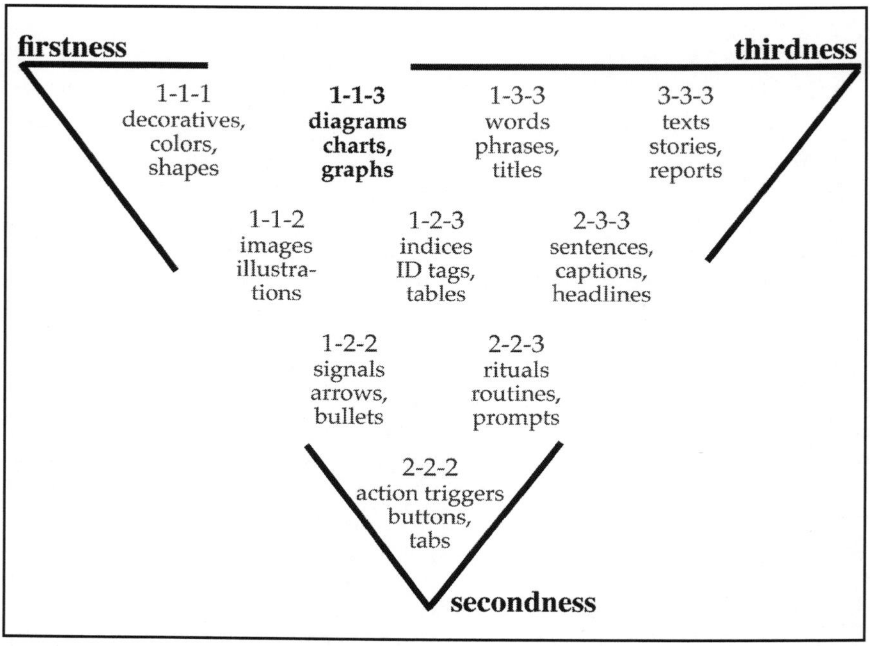

Figure 6. Diagrams (boldface) located relative to other visual types (muted).

Diagrams are the first intermediate step between image-like perception and knowledge expressed in language. Diagrams act as a bridge to convert perception into verbal information and vice versa: verbal information into perception.

- Diagrams properly configured as information-perception bridges share some properties with decoratives and images (i.e., they should be attractive and can somewhat resemble actual things), but at the same time, diagrams should share properties with words (written or spoken). That is, words are made up of distinct, contrastive parts (letters or sounds), each part is relevant to a word's meaning (i.e., no extra or superfluous parts), and word meanings focus on general ideas. Likewise, diagrams should be made up of distinct contrastive parts, each part relevant to the overall meaning and that meaning related to general ideas.

DIAGRAM APPLICATIONS AND LIMITATIONS

Diagrams (including visuals like maps or graphs) communicate statements or ideas that can be validated (judged true or false) by means of logical reasoning or experiment. Decorative-evoked feelings are not true or false; we have them, or we don't. Image-reflections of objects are not true or false; they simply exist, in the imagination if not in the physical world. Only informative statements like *Resin A formed an 80% stronger bond* (or it didn't), or *Cover plate B is secured by four hex screws* (or it isn't), or *Orlando is north of Miami*, etc. can be judged true or false in the usual sense. The informative content of diagrams can typically be paraphrased in terms of such statements.

Mediation between Language and Perception

Diagrams and other informative visuals (indexical tables and social signals discussed in chapter 5) mediate between perception/action (neither true or false) and the true-or-false statements of language (Figure 7).

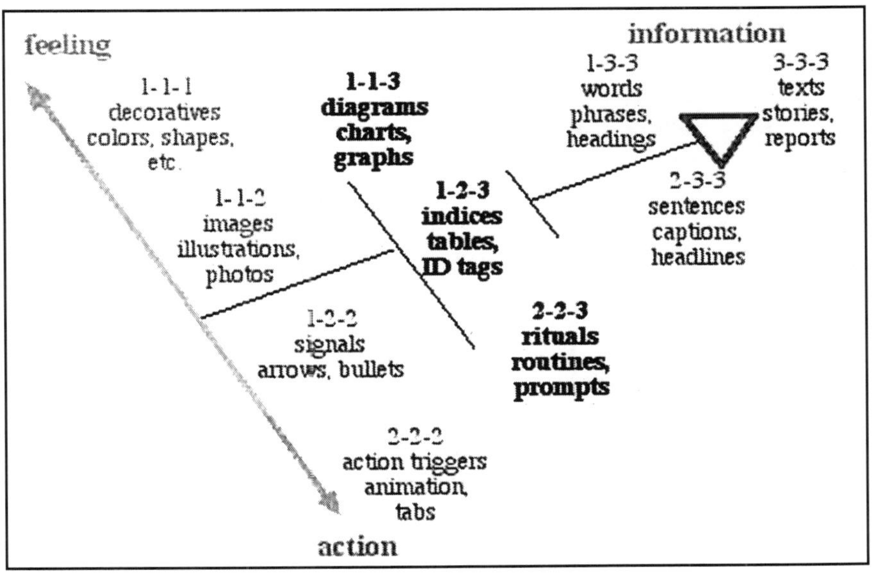

Figure 7. The mediating function of informative visuals: diagrams, indexical tables, and social signals.

Perception/action face of the Peircean pyramid (decoratives, images, signals, action) is connected to language and vice versa only through the mediation of diagrams, reference indices, and ritual-sequencing visuals.

As was mentioned in chapter 2, decorative/indicative images mediate between raw perception and raw action. How we react to a tickling feeling on the neck depends on how we imagine the source of the tickle, etc. In a similar fashion, diagrams and other informative visuals act as a bridge to convert verbal information into perception/action.

For example, the language statement *Digicom stock is about to lose 50% of its value*, if meaningful and true, ultimately has to translate into feelings (worry, if I own the stock; elation, if I sold it already) and action (i.e., to try to sell the stock or refrain from selling if the SEC will hang me for trading on insider information).

However, to move from language (Figure 7, top right) to the perception/action axis (Figure 7, left), information has to first pass through an intermediate stage (Figure 7, center), and that intermediate stage consists of informative mental visualization (in this case, something like a graph showing the steep fall in stock price). Peirce was emphatic on this point: the notion that language has to pass through intermediate visualization in order to be understood.

> In all primitive writing, such as the Egyptian hieroglyphics, there are icons of a non-logical kind, the ideographs. In the earliest form of speech, there probably was a large element of mimicry. But in all languages known, such representations have been replaced by conventional auditory signs. *These, however, are such that they can only be explained by icons.* But in the syntax of every language there are logical icons of the kind that are aided by conventional rules. (CP, 2.280, emphasis added)

Thus, in Peirce's system, the underlying meaning of any word, sentence, or any longer text (Figure 7, top right) must be understood as a diagram or other icon, either visually or in some other sensory mode, either consciously or unconsciously (Figure 7, center). The meaning of any verbal information, if it is understood, has to be translated into both perception (i.e., what we'd see, hear, or feel if the information is true) and action (i.e., how we'd act if the information were true). This, in theory, is the fundamental reason that we use visuals to present complex information: in textbooks, in assembly instructions, in maintenance manuals, in PowerPoint presentations, or anywhere else.

When information is complex, the labor of converting it into complex mental icons is high. Either a speaker can foist upon the audience all of this labor of reconstructing verbal information as visual or other sensory information, or a speaker can at least partially process the information visually, in advance, on behalf of the audience, by providing some informative visuals.

There is enormous temptation to shortcut this process of intermediate visualization by skipping straight from language to perception, straight from words to images. This temptation should be resisted. Images, particularly photographs, have a number of cumbersome properties if the primary goal is informative communication; in other words, if the visuals were chosen to support textual information. Recall the point made by Walsh in chapter 2, about a lecturer who went through hundreds of photographed images of Baroque architecture without ever effectively explaining how to recognize the Baroque style.

> The psychological problem created by the use of photographic images without sufficient orientation appears to be that non-specialists process a photograph of a building or painting holistically, as a unique entity. Without explicit instruction on what details are pertinent to the period, any details that they perceive are likely to be irrelevant. This sort of processing will result in an impressionistic, rather than systematic, comparison of successive photographs. This procedure is analogous to the classification of books by their size or color rather than their subject matter. (1998, p. 210; see Figure 8).

As Walsh notes, natural images tend to bury relevant detail because they lack differentiation between relevant and irrelevant detail. Consider for example some slide images of a the kind of Baroque architecture mentioned by Walsh in Figure 8, left, as opposed to the kind of diagram Walsh describes as being more effective, center. Important details of the architectural imagery blend rather seamlessly with details that are largely irrelevant.

To be grasped, the images have to be perceived as a holistic mass of object detail. And yet, to be understood, information has to be clearly divided into parts that strongly contrast. This is the primary difference between images and diagrams. Moreover, the contrastive element of diagrams can help the reader or viewer better retain the information of the visual.

> Diagrams and drawings, together called line art, can sometimes do what photography cannot. As the term suggests, line art usually employs lines or other abstract shapes to show something. It presents an abstracted version of what someone might see, stripping away all the extraneous details and concentrating the user's attention on the important details.
>
> Line art also allows us to show users things that a camera can't access—such as the inside of an object. (Miles & Hawkins, 2008, p. 214)

Signal-processing engineers understand that an information signal at its most basic level has at least two parts: a carrier and a modulator. Modulations contrast with the basic carrier. Programmers understand that computer code has to have both zeros and ones, both "on" and "off," which have to remain distinct. Writers understand that a sentence has to have a subject and predicate (i.e., a noun and a verb). Without distinct, contrasting parts, there is no signal, no code, no sentence, and no information. This contrastive property of information carries over in diagrams that mediate between information and perception.

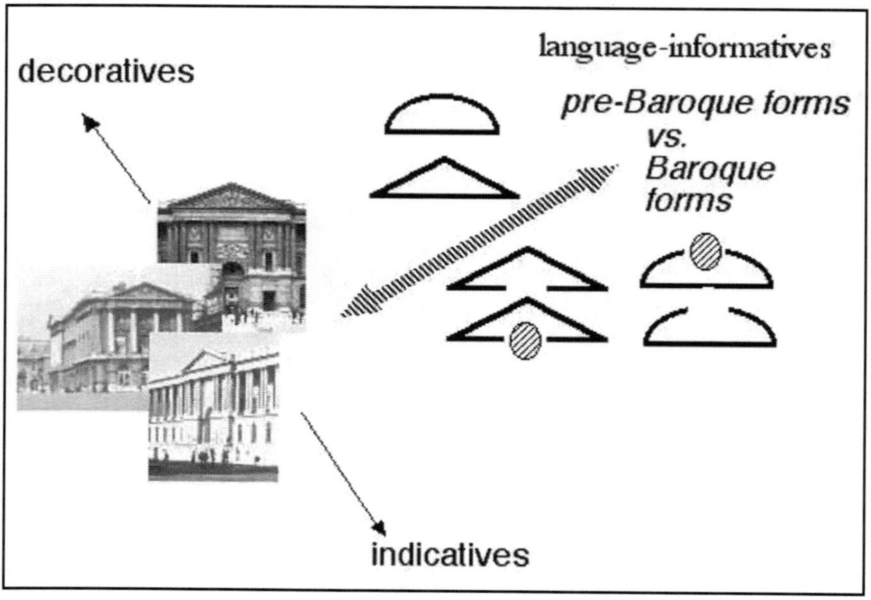

Figure 8. Architecture example images.
from www.arch.montana.edu/classes/arch323/French%20baroque.htm

Images communicate feeling and indicate specific objects; diagrams inform by emphasizing contrasts, restricting detail, and abstracting general propositions from an indefinitely large number of individual images/objects.

Because images inherently blur all such contrasts, they do not, in principle, provide information by themselves. As we explained in chapter 2, an image by itself can neither affirm nor deny a proposition. Changing the caption of an image can utterly change its message (as in Figure 2 in chapter 2). Effective diagrams likewise need significant verbal support but for a different reason. The diagram is clarifying the meaning of the verbal text. Diagrams, in order to count as diagrams, must have been constructed with particular conceptual relationships in mind, as articulated by language. Putting a different caption on an image gives it a different, but equally viable, meaning. Putting a different caption on a diagram can only clarify exactly the same meaning that it was meant to have, or else create a kind of visual "lie" if the diagram is thus mislabeled. Figure 9, for example, has only one effective interpretation, as the relationship of parts in a motorized valve.

The multiple possible interpretations of any given image can be used effectively to affect a wide audience with a strong but nonspecific emotional message for example, but multiple interpretations of any key aspect of a diagram invariably create confusion. This confusion can be due to insufficient prior training on

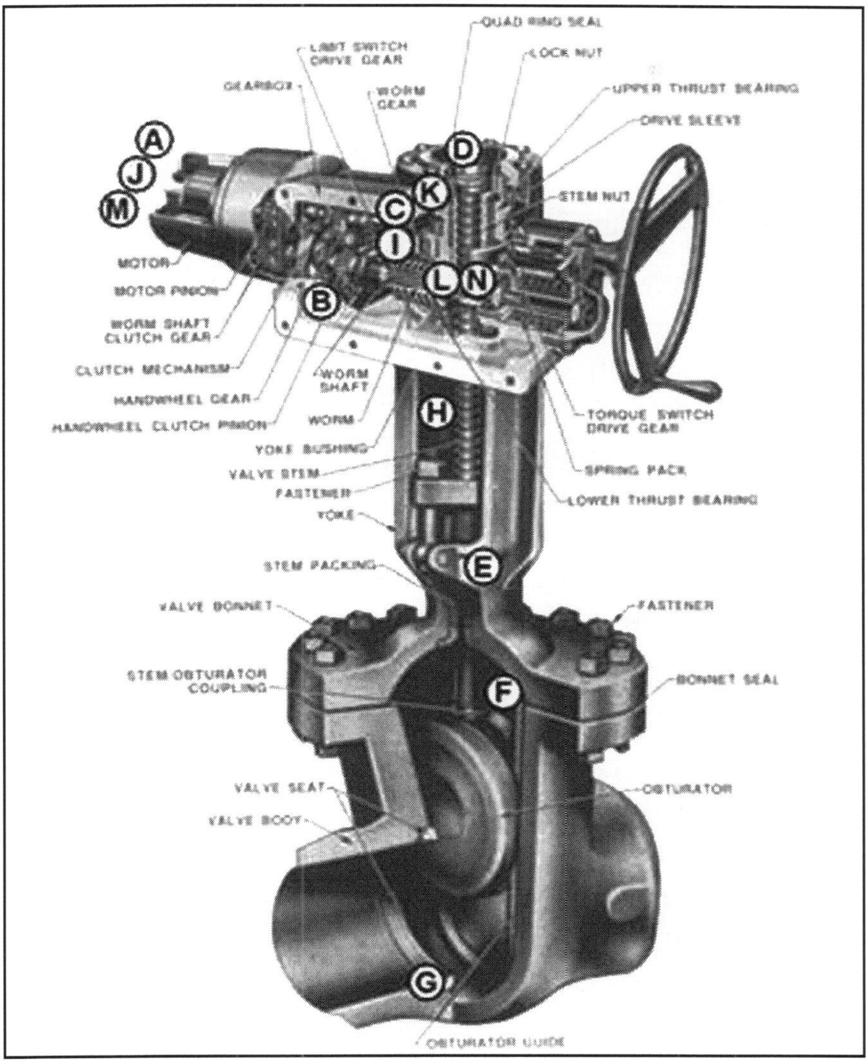

Figure 9. Single correct interpretation for diagrams.

A diagram usually has only one correct interpretation, even if that interpretation is not immediately clear. For example, a diagram of a motorized valve can only be exactly that, regardless of the caption.

the part of viewers, but it is also typically due to poor design or poor captioning of the visual.

Detail Management

A second critical difference between images and diagrams is that images correspond point-to-point with things in nature, and as such, images have no filter for irrelevant details. Whatever details are relevant to an informative purpose are easily lost in the crowd of details that images naturally contain. Again, compare the architecture images in Figure 8 with the diagram where only the critical features are represented.

Every single detail of a diagram should be evaluated to make sure it contributes part of the information conveyed in the diagram as a whole and that the information in the diagram matches actual audience need. Generally speaking, audiences have a goal of accessing information in a timely, efficient way. Unnecessary visual clutter in a diagram interferes with that goal and thus creates ethical difficulties (Manning & Amare, 2006). In most cases, both details and actual information in a diagram should be trimmed back to just the essential points, "essential" here being determined by a careful audience analysis. Figure 10 illustrates this kind of detail adjustment.

A well-formed diagram includes only the detail necessary for its purposes. For example, a detailed "wire" diagram is unnecessary to indicate the meaning of basic aircraft terms (Figure 10). Though they are often the superior design choice, diagrams are usually a more difficult choice for authors and speakers using PowerPoint-style slides, primarily because diagrams have to be deliberately constructed, whereas images can be created with the mere click of a camera button. The temptation to use easily created images is further compounded by the difficulty in creating effective diagrams, which are unquestionably "more laborious to design, and they take up more space on paper or on the computer screen" (Boekelder & Stëehouder, 1998, p. 231).

The difficulty in constructing effective diagrams is further complicated by an unavoidable difference between the knowledge base from which diagram designers work, all that their diagrams might represent, and the tiny subset of that knowledge base that their viewers/users need to know. Layers of detail with significance to diagram designers may only serve to distract less knowledgeable viewers. It's important to remember that if a diagram contains more information than viewers can actually process, it degenerates into a mere image of information. We've all experienced this when we "see" a page of text but don't actually read it—all we see is a block of black words on a white page or screen. In other words, we see the physical reflection of information that someone wrote, which is essentially just an image, until or unless we process the informative content.

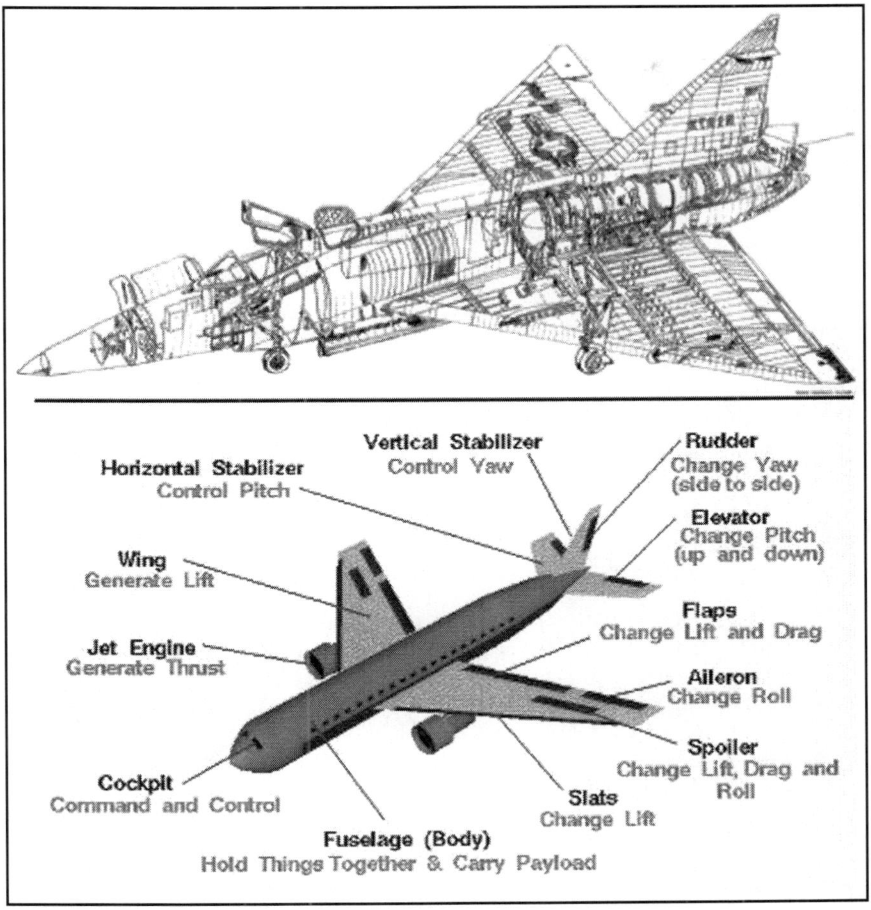

Figure 10. Overly detailed vs. stripped-down diagram.

A well-formed diagram includes only the detail necessary for its purposes. For example, a detailed "wire" diagram is unnecessary to indicate the meaning of basic aircraft terms.

Likewise, if a diagram is an overly complex block of lines and arrows and shapes, viewers can only *feel* that the visual is informative because it resembles other diagrams and physically indicates the presence of information. But if they cannot actually extract the information, then the visual, despite its diagram-like appearance, is for these viewers merely a decorative-indicative visual, that is, an image.

General Ideas

A third critical difference between images and diagrams is that images focus on traits of an individual case rather than the general properties of a group. It is difficult to draw reliable conclusions from the image of just one particular individual if the goal is to communicate the general properties of the type of thing the image is meant to represent.

Imagine, for example, the odd tactic of differentiating restrooms by placing the photograph of some specific woman on one door and the photograph of some specific man on the other door. The typical diagrammatic forms do a much better job of communicating the general idea. Likewise, a photograph of a cat is just a cat, but a cartoon cat (like Felix or Garfield) can take on general anthropomorphic properties and thus come to represent human foibles and wiles in general. For similar reasons, photographs of a specific device are almost never as informative as a general diagram showing the layout of all devices of that type (see Figure 11).

It's worth noting that Stan Lee, creator of some of the most successful comic book series (*Spiderman, The Incredible Hulk, Fantastic Four*), also developed instructional materials for the U.S. Army.

> When I was in the Army I wrote training films and I wrote training manuals. One of my jobs was to take subjects that the troops were having trouble learning and rewrite the manuals so they could absorb that knowledge faster. *And what I did was I used more illustrations. I came as close to the comic strip format as I could.* I also added a bit of humor, which helps tremendously. And we were able to speed up the training processes in a lot of areas. (Ewalt, 2005; emphasis added)

As was noted earlier in this chapter, comics and cartoons are the first step in the direction of informative diagramming. Comics and cartoons likewise step away from imagistic levels of detail, which allows for clearer contrasts and more reliable generalizations to be communicated. McCloud (1994) likewise recommends comics as an effective instructional medium, superior to images for conveying general ideas.

CONCLUSION: TOWARD A DIAGRAM ETHIC

Ethical improvements to informative communication generally depend on shifting our orientation away from an image-driven orientation in visual design. Instead, we need to move consciously toward diagrammatic modes of presentation, always keeping in mind these essential distinctions (see Table 2).

Unfortunately, the distinction between images and diagrams is not well or widely understood, and this lack of understanding can lead to ethical difficulties, as we have already seen in previous chapters—a disjunction between the author's

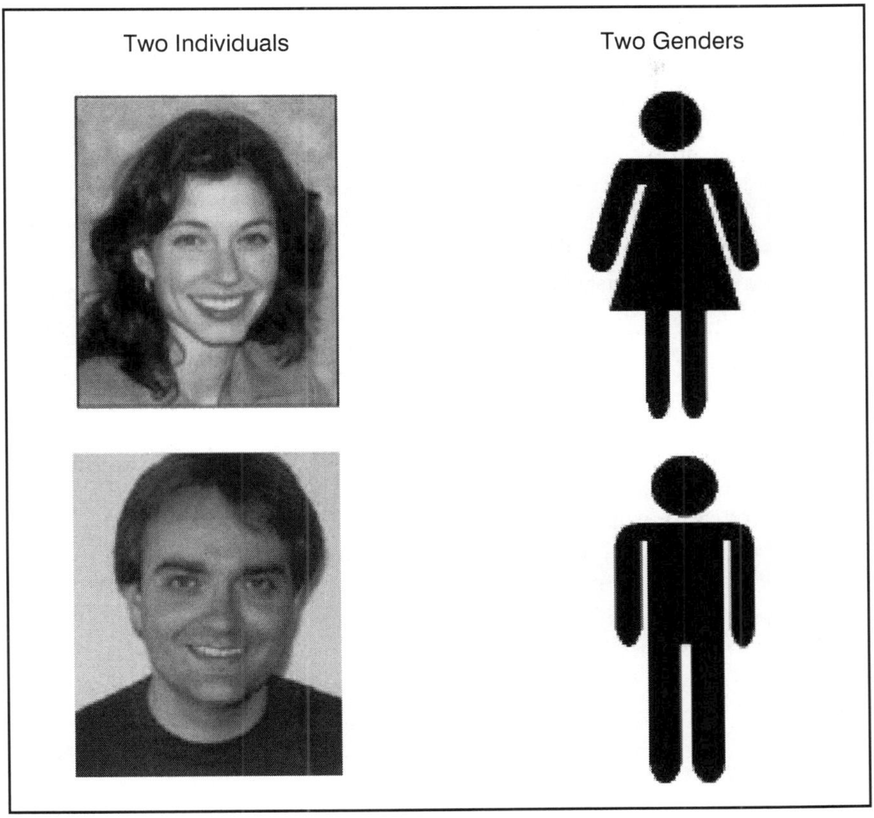

Figure 11. Photograph vs. diagram: one object vs. the general idea.

Typical diagrammatic forms do a much better job than images communicating a general idea.

informative responsibility to an audience, usually best served by diagrams, and an author's personal desire to make a visual decoratively interesting and indicative of their visual design skills, a covert and ethically dubious goal that is nevertheless best served by images. What makes the task of ethical improvement to visuals even more difficult is the institutionalization and commercialization of a self-serving image-driven ethic in corporate culture:

> Why just get by with a boring presentation when you can create an engaging Microsoft Office PowerPoint presentation or a colorful Microsoft Office Visio diagram? Get ready to impress the big boss or the new team with simple ideas that go a long way. (Microsoft, 2003)

Table 2. Essential Distinctions between Images and Diagrams Summarized. Visuals Intended to Convey Information (i.e., Diagrams) Should Have Properties Distinct from Visuals Intended to Indicate Objects and/or Evoke Feeling (i.e., Images).

Images:	Diagrams:
1. Lack of clear contrasts	1. Include clear contrasts
2. Mix relevant and irrelevant details	2. Filter out all but relevant details
3. Represent specifics better than generalizations	3. Represent generalizations better than specifics

The emphatic message here is that simple ideas are not good enough and that the essential information needs to be padded out with decorative forms and ornamental imagery to impress both bosses and co-workers. Though pervasive, this message can be successfully resisted if we can consistently demonstrate through well-constructed diagrammatic presentations that there is an even more impressive mode of communication that is both clear and informative.

In subsequent chapters, we will show how the essential visual form of diagrams is reflected in effective tables, indexes, ritual sequences, and language itself, all with minimized detail and emphasized contrasts to be interpreted as general propositions.

CHAPTER 4

Indicatives

Traditional visual design has not made an effective distinction between visual "contrasts" that are merely decorative (i.e., *varieties* of shape, space, line, color, texture, and typeface that create aesthetically "interesting" or "attractive" visuals) and those sharper visual contrasts that provoke physical action in viewers. In Peircean analysis, this is a critical distinction. In particular, strong lines, bullets, arrows, line-enclosed negative space, or radical changes in size, color, typeface, etc. all serve the primary function of physically directing audience focus. Any amount of decorative variety may be acceptable as long as the whole remains unified, but these strongly contrasting, strongly indicative elements can only be effective if used sparingly, in measured doses in any given figure or page. The Peircean analysis explains exactly why indicative contrast is irritating and anxiety producing if overused. We classify as *indicatives* those visuals that primarily provoke action. The exact nature of that action differentiates the two major indicative subtypes, namely *signals* and *action triggers*.

- Bullet points (and the white space between bullet points) are typical *signal* elements, as are arrows, boldface/italic/underlined text, and contrastive colors. These all can serve the primary function of directing viewers to focus eyes and minds on a specific location adjacent to the signal itself.
- Web links, buttons, and page tabs are typical *action triggers*. These all have two parts: (a) a signal that draws visual attention but also (b) a *physical cause-effect chain* resulting from the physical actions of moving a cursor or finger over the signal, then clicking, pressing, flipping, etc.

Note that animated or blinking signals have physically stronger effects on viewers than static signals, and these moving forms technically should be classified as action triggers because they likewise consist of cause-effect chains of distinct forms (the chain of pictures or forms linked together to make the animation).

Whether they are static signals or dynamic action triggers, the common property of indicative visual forms is that they provoke action or trigger chains of action, at minimum the act of focusing attention. For this very reason, any signals or action triggers must be used carefully, in measured doses on any given slide, display, or on a Web page. To do otherwise creates an exhausting welter of visual activity (Figure 1).

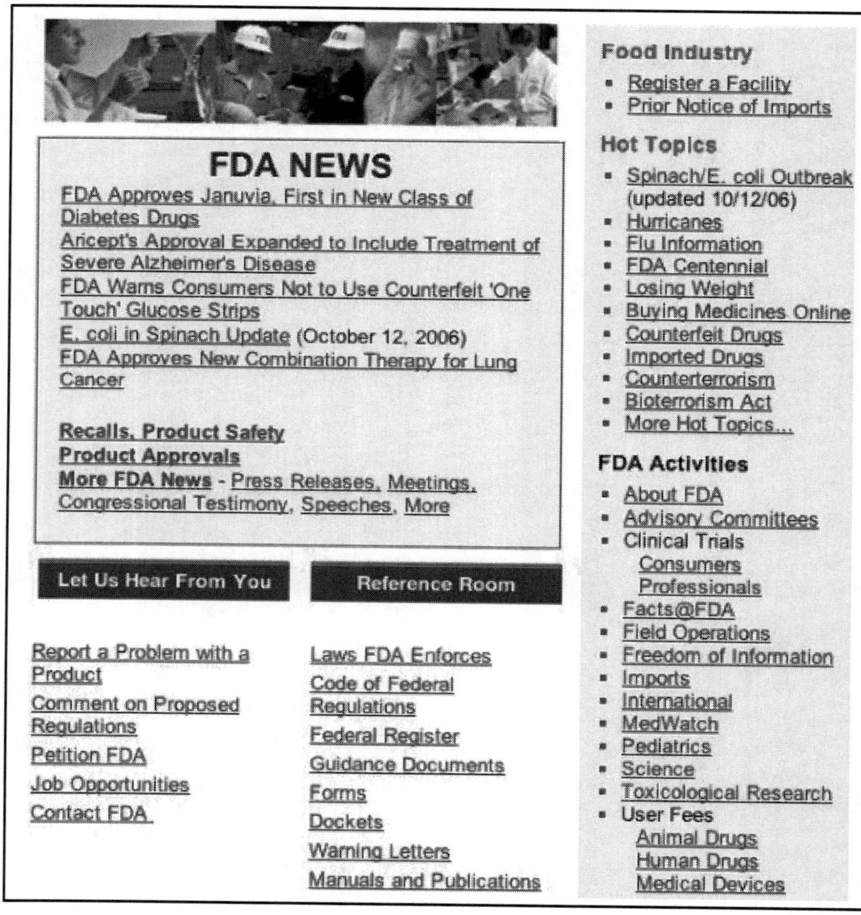

Figure 1. fda.gov (2006), a dense thicket of bullets and links.

Unorganized lists of links impose undue levels of effort on viewers attempting to locate specific information.

To overuse either of these indicative modes, either too many signals (e.g., bullet points) or too many action triggers (e.g., links) in one visual display, is to overwork users, both mentally and physically. Fatigue sets in, important points are overlooked, and the visual fails in its purpose. To drag users through a series of actions without purpose would also have clearly negative ethical implications.

We want to begin by identifying those relatively pure indicatives, those that point to adjacent informative content (the focus of this chapter). These pure forms contrast with indicatives that are structurally merged with informative content (i.e., ordered labels, tables, indexical lists, and ritual action sequences, the focus of chapter 5). These distinctions are diagrammed in Figure 2.

In Peirce's model of sign types, there is a logical progression from the single triangle of icon subtypes (decoratives, images, and diagrams) to the "double triangle" consisting of two pairs of index subtypes. Just as images are "elevated" into diagram forms, so too, are signals and action triggers "elevated" into reference indices and ritual sequences. This parallel elevation of two separate types—signals and actions—creates four indicative forms altogether, or in other

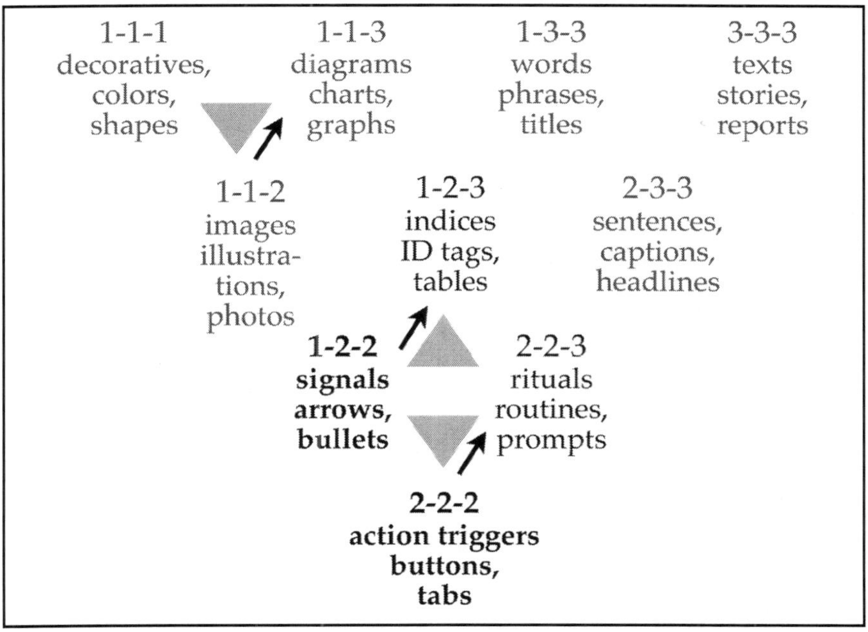

Figure 2. Pure signals and actions, as opposed to informative indicatives.

Just as images are "elevated" into diagram forms, so too, are signals and action triggers "elevated" into reference indices and ritual sequences.

words, a "double triangle." Furthermore, decoratives, images, and diagrams, as icons, all holistically resemble whatever is represented. Each of the four index types, in contrast, has a visible duality, a pointing signal plus what it points to.

Each of the four indicative subtypes must actively point to something outside of itself. Put another way, each indicative type of visual must physically connect to something outside of itself and call *more* attention to whatever it points to, considerably more than the attention it calls to itself. For example, underlining serves not to point the reader to the line itself but rather to the word above the line.

The core example is an arrow glyph: →. If we ignore whatever the arrow points to, the glyph by itself is iconic, if taken in solitary form. However, the normal use of the arrow glyph is not to represent what it resembles (e.g., a projectile weapon or perhaps a bird's footprint). Rather, the arrow represents a call to action, to notice whatever is physically next to and/or in line with the arrow.

The three kinds of icons are →

→ decoratives
→ images
→ diagrams.

Whether it looks like an arrow or a star, a colon, dash, or dot, we find here the essential indicative meaning of any bullet point as an indicative signal: the signal draws attention, that is, mental focus that is a kind of perceptual action, but it directs that attention to something other than the signal itself, most usually the text or figure next to it. We can accomplish the same effect by surrounding each line with white space, instead of or in addition to marking each line with a bullet character. An overt bullet point, like an arrow or star, is not often strictly necessary, except to explicitly signal list items as a group on a single indicative line. In the absence of a stronger indicative purpose, the white space by itself may call adequate attention to what is next to it, calling little or no attention to itself.

The three kinds of icons are

- decoratives
- images
- diagrams

See-but-look-elsewhere duality is the hallmark of indicatives as opposed to the *look-here* unity of decorative forms, which primarily call attention to themselves. The indicative signal provokes the viewer to notice and divide visual elements that are distinct from the signals themselves. Using a dash or hyphen for a signal instead of an arrow or white space, we find the same effect. The four kinds of indicatives are

- signals (Peircean type 1-2-2; e.g., bullets)
- action triggers (type 2-2-2; e.g., click link→load page)
- reference indices (type 1-2-3; e.g., tables) and
- ritual sequences (type 2-2-3; e.g., knock-knock→who's-there?).

This chapter will focus on signals such as arrows or bullet points, and action triggers such as Web page links, buttons, and tabs. We will also discuss animation in the form of blinking or moving signals, which are often used just as if they were ordinary signals, but which are inevitably covert action triggers, their movement literally forcing viewers to look at them.

Signals and action triggers are both common in visual display, and in either case, these provoke an active physical response in audience members. Signals cause a viewer to separate mentally or to contrast whatever bits of information they point to. Action triggers require viewers to physically follow, manipulate, drag, mouse-over, or click some visual display or hypertext. In the right context, these are effective tools of visual information design, but because they impose actual mental or physical effort on users, indicatives should be used only with *measured*, deliberate purpose, not indiscriminately and with little thought given to the workload imposed on users.

The term *measured* will itself require some discussion and clarification. Essential points of information can typically be communicated with 3 or 4 indicative bullet points per page or per screen rather than 15 or 20 points per page or per screen. Deploying only the minimum necessary number of indicatives means that audience members are asked to expend only the labor that is necessary to interpret a text (as illustrated, for example, in Figure 3).

Experimental evidence suggests that even an undifferentiated list of seven bullets or Web page links can be excessive, causing viewers to overlook or forget key points. This was clear even in the experimental data reviewed by Miller (1956; his oft-cited but not well-understood "Magical Number Seven Plus or Minus Two" article). In a typical experiment, tones were assigned numbers, and listeners were asked to remember which number went with which tone. "When only two or three tones were used, the listeners never confused them. With four different tones confusions were quite rare, but with five or more tones confusions were frequent. With fourteen different tones the listeners made many mistakes" (p. 83).

In this and several other experiments, replicated in recent research (Cowan, 2001), four is identified as the practical limit of unorganized items that people can *reliably* hold in memory, while five to nine items (seven plus or minus two) remains the upper limit or "channel capacity," or in other words, *the point at which memory performance becomes unreliable.*

Thus, seven-digit phone numbers are consistently divided into one chunk of three and one chunk of four—555-1234—this division to facilitate easier recall.

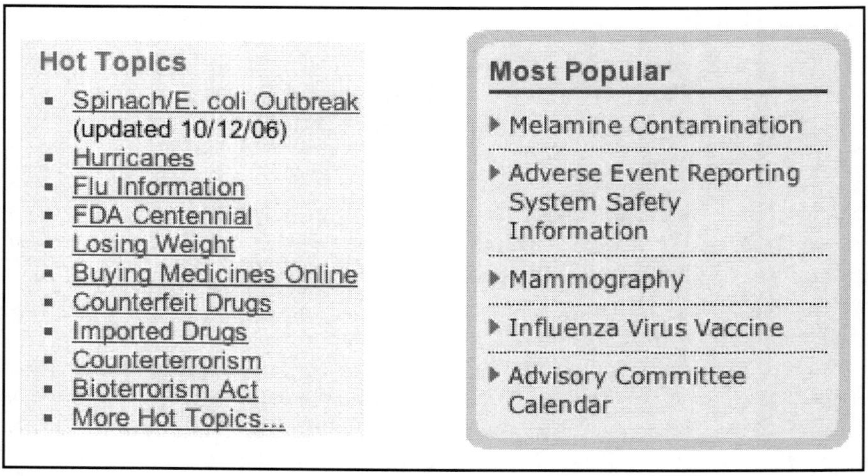

Figure 3. fda.gov (2006 vs. 2008), with relatively fewer bullets, more effective than many.

Bullet points (and indicatives generally) should be used only with measured, deliberate purpose, not indiscriminately and with little thought given to the workload imposed on users.

Bullet lists, Web page links, and verbal instruction steps (etc.) all require even more of memory and visual tracking than the simple recall of individual numerals. There is correspondingly an even greater need to visually separate and divide any lists containing more than four or five items into smaller, more manageable chunks. Even so, it is utterly common to find undifferentiated lists of seven, nine, or even more items that unquestionably cause viewers to overlook, forget, or confuse some of the items.

In spite of these well-documented facts, a fairly tenacious myth circulates in visual-design circles that seven-item or even nine-item lists are fully acceptable.

> What George Miller presented as "some limits on our capacity for processing information" [p. 81] quickly turned into an acceptable average: Miller's whole paper is now ignorantly summed up as "seven is OK": seven items in a bulleted list are readable, seven main points in an oral presentation are fine, seven levels of heading in a written document are acceptable. This proposition is what Steven Pinker would call a CONVENTIONAL ABSURDITY: "a statement that goes against all common sense but that everyone believes because they dimly recall having heard it somewhere and because it is so pregnant with implications." (Doumont, 2002, p. 123; emphasis in original)

Given the pervasiveness of the seven-is-OK myth, we again must emphasize here that a visual display with seven otherwise unorganized indicative points (anything more than five really) imposes too much active labor on users, unless there is some organizational chunking, as shown in Figure 4.

In the original Transportation Safety Administration (TSA) graphic, 20 layers of airport security are listed as separately indicated, separately colored bars (Figure 4). Using many separate indicatives induces far more visual activity than is required by the real message of the visual: *A terrorist attempting to bypass one layer of security will likely be discovered at another layer.* A revised version of the graphic (Figure 5) can be simpler, deliver the same message, and actually communicate more about the logic behind the organization of TSA security.

The essential message here is that indicative features such as bullet lists obey the same principles as other indicative-contrastive strategies, such as indicative color, boldface typefaces, blinking arrows, or any visual form intended to point to something beyond the form itself. The mind and eye can only process so many of these pointing elements at once (three or four at a time, research indicates) before they lose their effectiveness. Peircean analysis suggests an explanation: all indicative features require physical effort from viewers, effort that makes demands on a finite capacity for memory and action.

INDICATIVE DESIGN—PEIRCEAN PARAMETERS

Recall that each visual type has three key parameters, based on the Peircean categories:

- its basic form—whether the visual is unified or has distinct phases
- its referential process—how the visual acquires its meaning
- its interpretation type—what kind of meaning the visual form typically acquires.

Basic Form

This is the first chapter in which we consider two distinct types of visual sign simultaneously. In shorthand terms, these are types 1-2-2 and 2-2-2. Signal indicatives (type 1-2-2) and action-trigger indicatives (type 2-2-2) are alike in terms of referential process (physical adjacency) and interpretation type (physical action), but signal indicatives (such as bullet points) differ from action indicatives (such as Web page links) mainly in their basic form.

Signals are essentially static in form, meaning, for example, that the bullet points above, combined with the text they indicate, still form static, unified gestalts. In this way, they are not unlike iconic forms. In fact, signals (such as

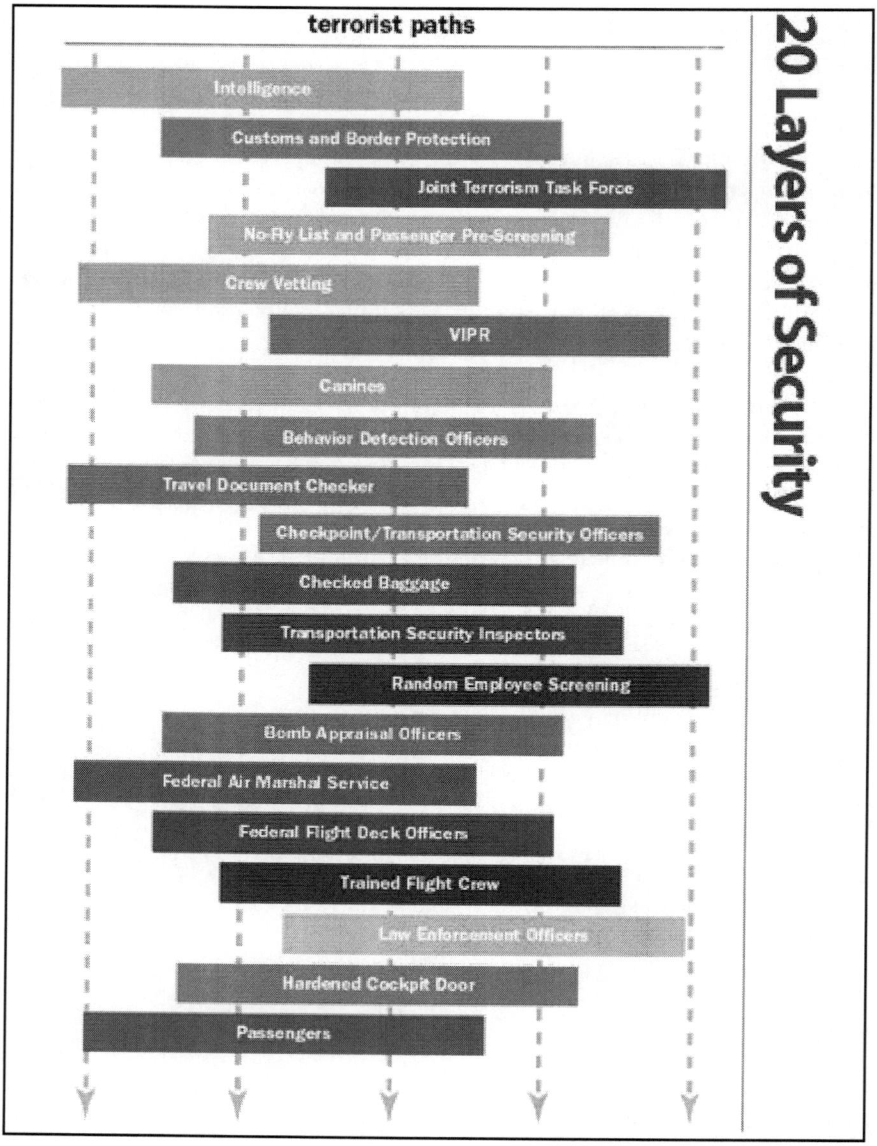

Figure 4. Unorganized indicative elements.
www.tsa.gov/approach/layered_strategy.shtm

Color, as an indicative, should be carefully measured in its use. Too many distinct color signals creates an overly busy visual, too much for viewers to process and remember at once.

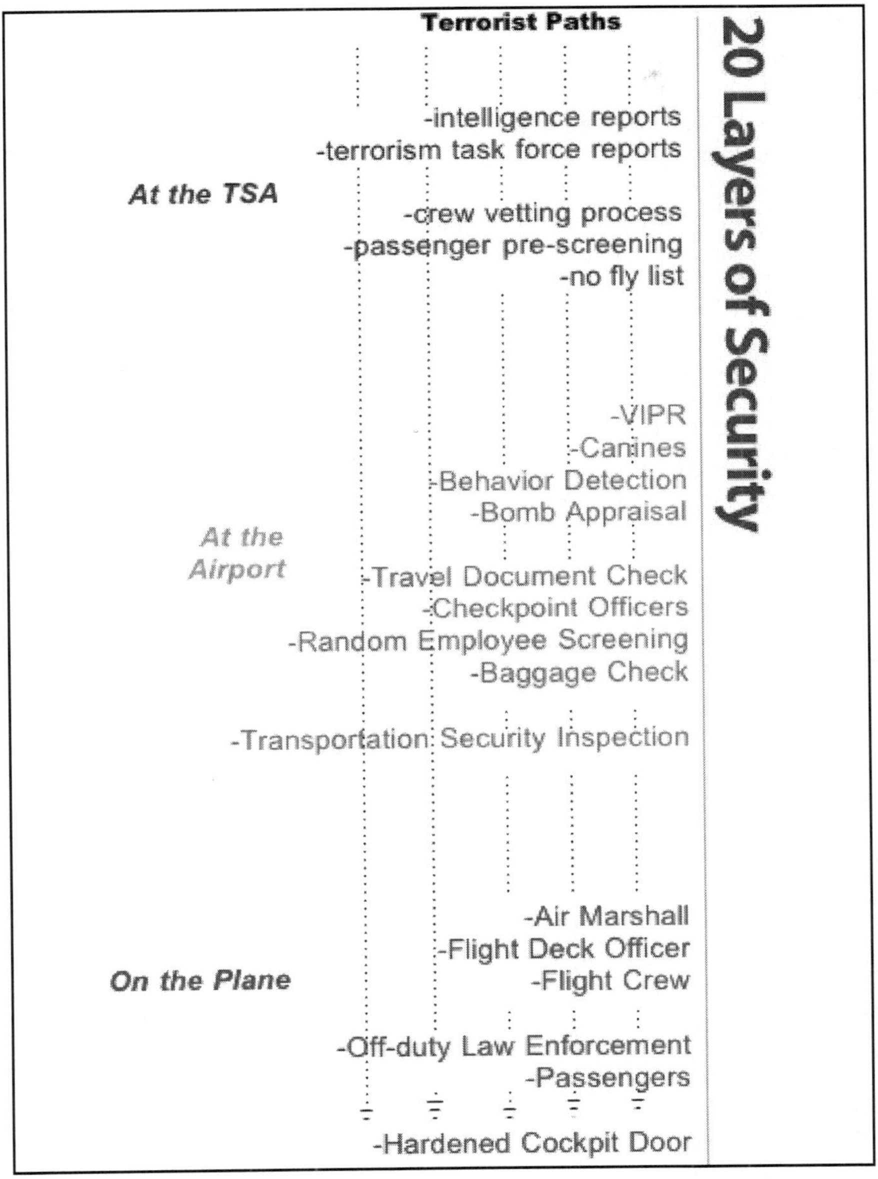

Figure 5. Organized indicative elements.

With less color, and organized indicatives, a revised version of Figure 4 can be simpler, communicate the same message, and actually communicate more about the logic behind the organization of TSA security.

arrows or stars) often would be iconic forms only, except for the fact that they call attention to something besides themselves, for example, as a cue for an upcoming list. Another indication that bullet points are something other than icons, even though they can look like icons, is that their particular form doesn't necessarily matter. As Schriver (1997) notes, a bullet is a "typographic cue that signals a list; it can be a dot, square, triangle, or other symbol" (p. 504).

Action triggers are like signals, and indeed they essentially incorporate some signal element. The incorporated signal can also be potentially icon-like, such as a tab or a button, but action triggers are essentially nonstatic in form, meaning that something dynamic happens when the action-trigger indicative is activated (as, for example, in Figure 6). A mouse-over brings up a menu; a mouse click brings up a new page or window or frame; a tab reveals a new layer of information, and so on.

In sum, the formal unity of any signal contrasts with the formal duality of action triggers. Compare Tables 1 and 2.

Signals provoke visual and mental action: separating, dividing, and contrasting otherwise static visual elements. Action triggers also visually separate, divide, and contrast—doing all that indicative signals might do—but action triggers rise through the level of visual unity to further engage physical action. The user must be able to physically follow, touch, click, or otherwise activate the link, tab, button, etc., the effect being in one way or another a dynamic change in the visual. Dynamic change is also the very thing that defines animation of any kind, which is why animation is technically an action trigger in terms of Peircean classification.

Referential Process

Apart from these formal differences, signals and action-trigger indicatives are otherwise alike. The process by which they acquire meaning is invariably by physical connection (as opposed to iconic similarity, where, for example, the arrow signal only resembles arrows incidentally). In order to function, signals must be physically located (in other words, *collocated*) near, next to, or in line with whatever they are supposed to call attention to as in Figure 7. Action triggers must be likewise physically connected to whatever sequence of mechanical actions they trigger.

Indicative design choices can therefore misfire if a signaling element—an arrow, label, or tag—is physically misplaced and fails to connect with whatever it points to (Figure 7, top). Likewise, there is failure if an action-trigger link does not bring up the expected information, either by being "dead" or "broken," failing to "go anywhere," "do anything," or worse, if the link connects to a page or pop-up screen that is other than what is suggested by the link labeling.

Indicative design choices can also fail if the indicative signal or action trigger calls excessive attention to itself, becoming itself a distraction, such that the

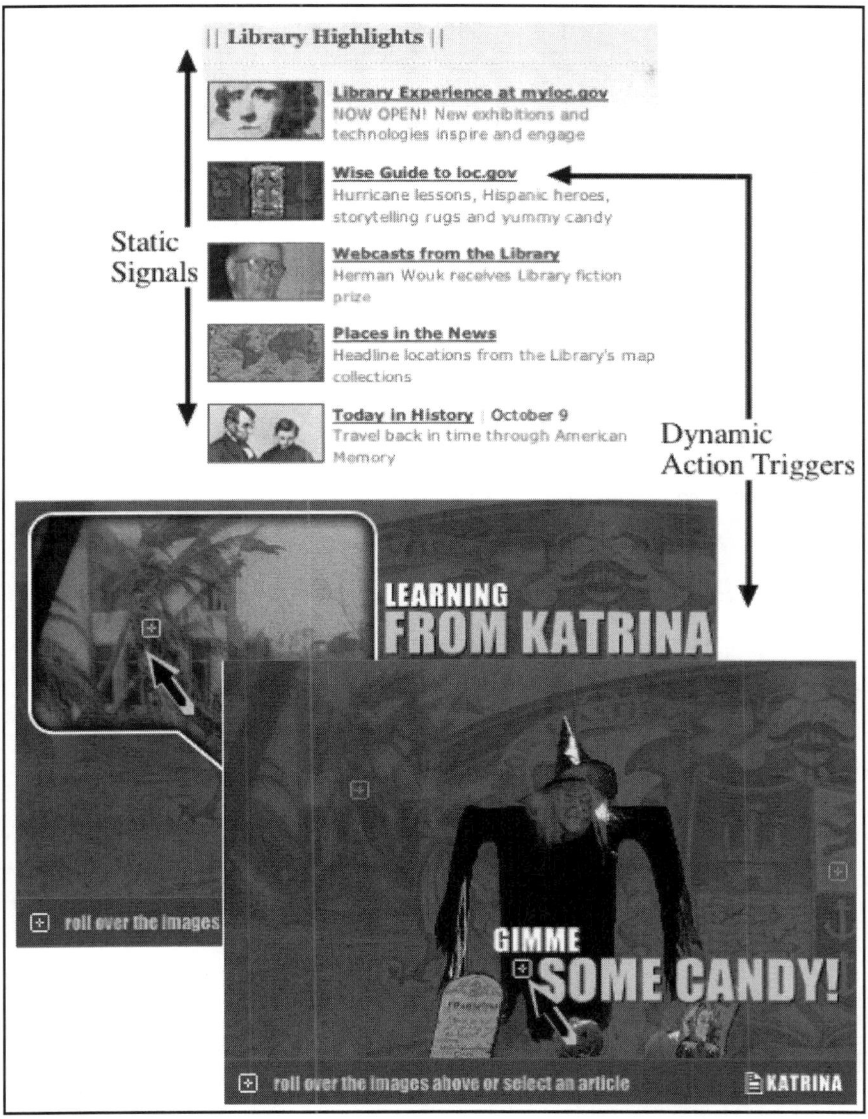

Figure 6. Static signals vs. dynamic action triggers.

Action triggers consist of a visual signal plus a dynamic visual change when they are activated.

Table 1. Signal Indicatives, Shorthand Code = 1-2-2[a]

Basic Form	Referential Process	Interpretation Type
Qualities of unity	Qualities of adjacency	Physical object connection
Form Level 1	Reference Level 2	Interpretation Level 2
1	**2**	**2**

[a]Signals are defined by formal unity (level-1 form), physical adjacency (level-2 reference), and physical object connection (level-2 interpretation).

Table 2. Action Triggers, Shorthand Code = 2-2-2[a]

Basic Form	Referential Process	Interpretation Type
Dynamic sequence	Physical adjacency	Physical object connection
Form Level 2	Reference Level 2	Interpretation Level 2
2	**2**	**2**

[a]Action triggers are defined by dynamic sequence (level-2 form), physical adjacency (level-2 reference), and physical object connection (level-2 interpretation). They are, therefore, the purest indicative type, consisting of raw Peircean Secondness (action-reaction).

thing that the signal points to is easily overlooked (Figure 7, bottom). This is particularly a risk when a signal is constructed using images or bright colors or distinct typefaces, which are all strongly decorative also and therefore tend to call excessive attention to themselves.

Interpretation Type

Signals and action triggers also share an overall interpretive goal in that they all physically connect to concrete objects, by placement or by active movement toward whatever object they indicate, invariably a separate, tangible thing that is pointed to.

In Peircean terms, the design parameters of signals are First Level in terms of form (unity), but action triggers are Second Level (dynamic sequence). Both types have the same process of reference (by physical adjacency, collocation, or connection), and the same kind of physical interpretation (as an active connection to an object). The interpreted active connection to an object can be mental, such as when we associate an image with that object that it resembles,

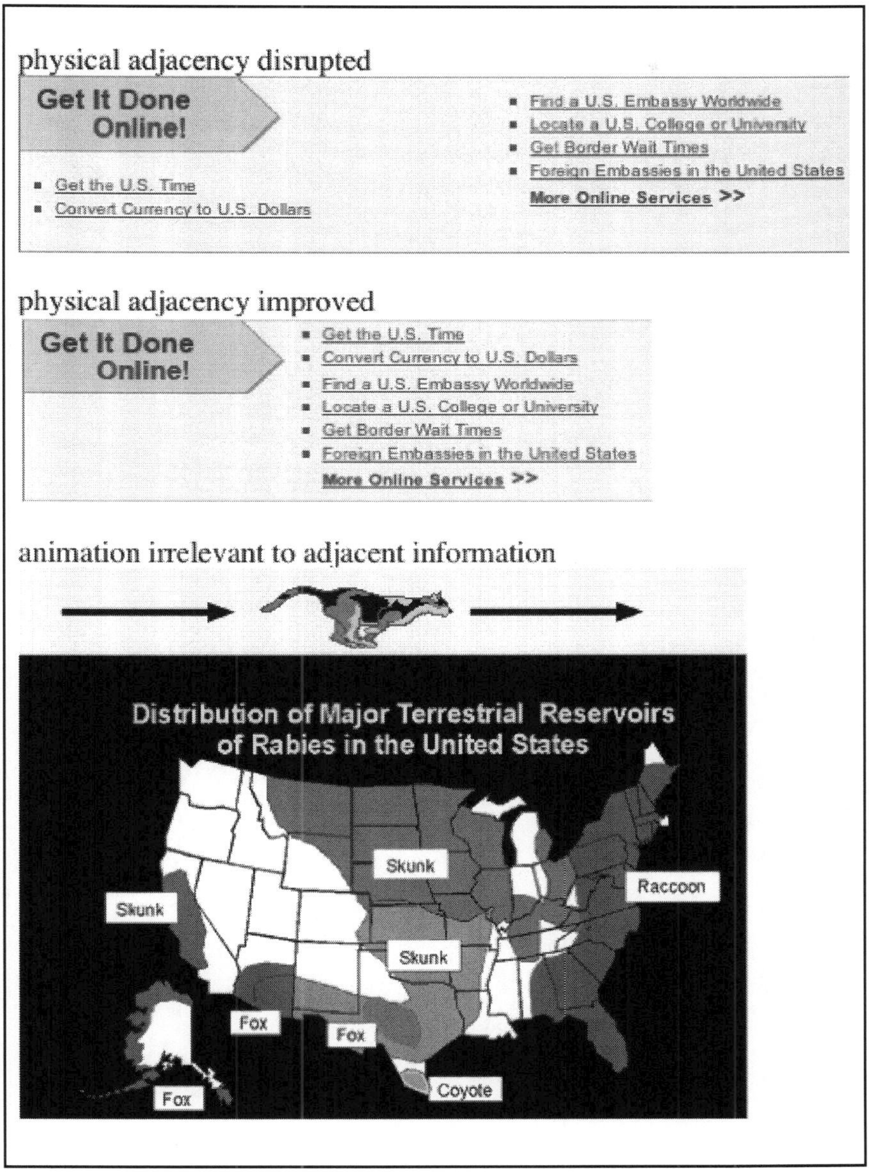

Figure 7. Reference by physical adjacency.

Indicatives are interpreted in terms of whatever they are physically adjacent to.

or with the content of a link that the image is physically placed next to. At the farther extreme of Peircean Secondness, that interpreted connection to an object can be a more visibly active touching, such as when we click the "buy" button at on online store, and this initiates a very physical chain of events culminating with the physical delivery of a product to our door.

Thus, in our Peircean shorthand, we refer to signals as 1-2-2 and action triggers as 2-2-2. This shorthand code locates these indicative types relative to other visual types in Peirce's system of signs (Figure 8).

Recall that the Peircean system shows communication development, starting from the top left corner of the inverted pyramid but progressing both downward and rightward. Therefore, the location of signals in the 1-2-2 position and action triggers in the 2-2-2 position (downward and rightward from decoratives and images, reaching the extreme lower point of the system) signifies the key properties of signals and action triggers as indicative types of visual communication.

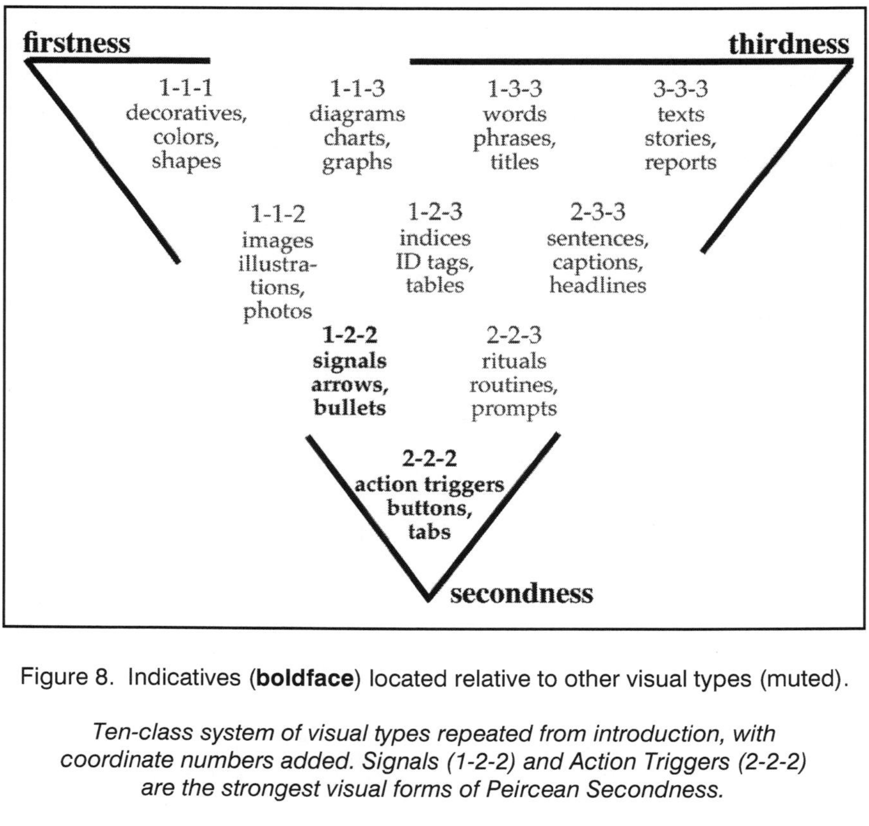

Figure 8. Indicatives (**boldface**) located relative to other visual types (muted).

Ten-class system of visual types repeated from introduction, with coordinate numbers added. Signals (1-2-2) and Action Triggers (2-2-2) are the strongest visual forms of Peircean Secondness.

- Signals, along with images, are intermediate steps between raw perception and raw action. Signals often incorporate images, such as the arrow bullets or page-tab motifs discussed earlier, but also the more detailed and provocative images used in sales pitches or Web pop-up screens. These image-signal combinations then induce viewers to immediately act, to enter a store or click on a Web link to a commercial site, etc.
- Action triggers, such as the pop-up Web page links just described, invariably incorporate a signal, so viewers can see where the trigger is. The signal in turn typically incorporates an image-like form, and the image typically incorporates decorative forms. In other words, the whole "leading edge" of the Peircean visual system—the left edge of the inverted pyramid—is typically involved in a kind of cascade, from immediate, decorative perception to a dynamic, indicative physical action.
- Indicative signals and action triggers have powerful effects, and therefore there is strong temptation to misuse or overuse them, with ethically questionable results. The easy slide from raw decorative perception to raw indicative action can effectively bypass any real information or thought, as represented by "higher" visual forms in Figure 9.

This decorative-feeling-to-indicative-action cascade is leveraged to an extreme degree in addictive processes such as gambling and pornography, and to a lesser degree in dubious marketing pitches. The aim in each case is to engage strong feelings (usually intense desire) in order to provoke specific actions (usually to visit further sites, look at more advertising, and eventually hand over a credit card number). As much as possible, these addiction-driven businesses want potential customers to bypass any rational thought that might prohibit those actions. It is no accident, therefore, that the visual designs employed by these businesses are intensely decorative (to evoke feeling) and indicative (to provoke action) at the expense of genuinely informative content (which would encourage rational thought).

Visual designers whose main goal is to provide informative content, in contrast, should avoid the kinds of tactics typical of addiction-driven marketing. To use decorative and indicative strategies more ethically, it is important to moderate their raw effects, both by reducing the sheer number of decorative or indicative visual elements and by mixing them appropriately with informative elements.

INDICATIVE APPLICATIONS AND LIMITATIONS

As shown in Figure 9, decorative elements can be successfully used *in conjunction with* indicative elements, especially if the intent is to bypass information and thought. However, it is more difficult to successfully use *as indicatives* those visual elements primarily decorative in function, such as colors and typeface choices.

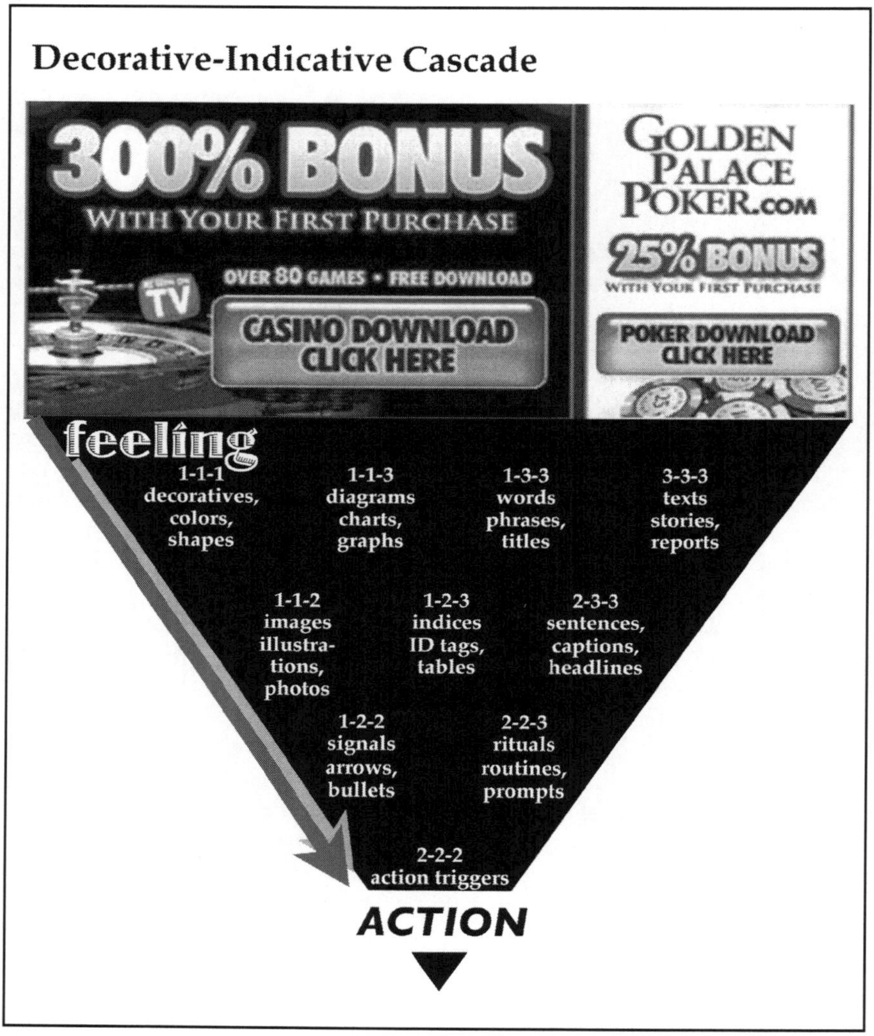

Figure 9. Dynamic cascade of visual types on the "leading edge" of raw perception-action.

If decorative and indicative visual strategies are emphasized, information and thought may be almost completely bypassed.

Decorative-Indicative Difficulties

We touched on the problem of using decorative forms for indicative purposes briefly in the opening discussion of decoratives in chapter 1. As shown by the removal test, Figures 1 and 2, it's possible for colors and typefaces to be used indicatively, but the immediate difficulty is that the colors and typeface forms tend to call more attention to themselves than to the separate object or information supposedly pointed to. Figure 1 is colored so as to resemble more than anything else an enormous bonfire, a mixed metaphor at best for a representation of the relative number of hurricanes and tropical storms. A similar problem was seen in Figure 7a (the colored cell-membrane diagram); again, the contrasting colors call more attention to themselves than to the distinct cell-membrane elements that the colors indicate.

A further difficulty using decorative forms as indicatives is that decoratives (code 1-1-1) require formal unity, (i.e., Peircean Firstness), but signals and action triggers (code 1-2-2 and 2-2-2) require a fair degree of contrast (i.e., Peircean Secondness), contrast that enables the indicative forms to stand out and get active attention but that must at the same time pass that attention to an adjacent object. Figure 10 further illustrates these difficulties.

The colored button links in Figure 10 strongly contrast with each other, which might seem to make them useful as signals/action triggers, but from a decorative standpoint, the colors both clash and call more attention to themselves than they do to the link labels. Likewise, the indicative title for this Web site, in a very conflicted red script typeface (amusing/agitating, see chapter 1), calls more attention to itself than to the attached image (Mt. Nebo, the dominant landmark in Juab County, Utah).

In terms of the Peircean visual grammar, a feeling does not imply action (a first thing doesn't imply a second thing), though action can imply a prior feeling. Likewise, there are no difficulties integrating words with indicatives because information (Peircean Thirdness) can imply action (Peircean Secondness), just as action implies feeling (Firstness).

In practical terms, this grammatical rule means that we can't effectively press raw decorative elements (like bright color or typeface forms) into service as indicatives because the first decorative effect of color or form will not and cannot by itself support a second indicative function.

We can however integrate forms that are separately indicative (such as arrows and lines) with decorative forms (such as color or typeface) on two conditions:

- the decorative does not call undue attention to itself and
- the feeling that the decorative form tends to evoke is not in conflict with the action we intend the indicative to provoke.

In Figure 9, for example (the decorative-indicative cascade of gambling casino visuals), the electric colors and bold block typeface are integrated with the

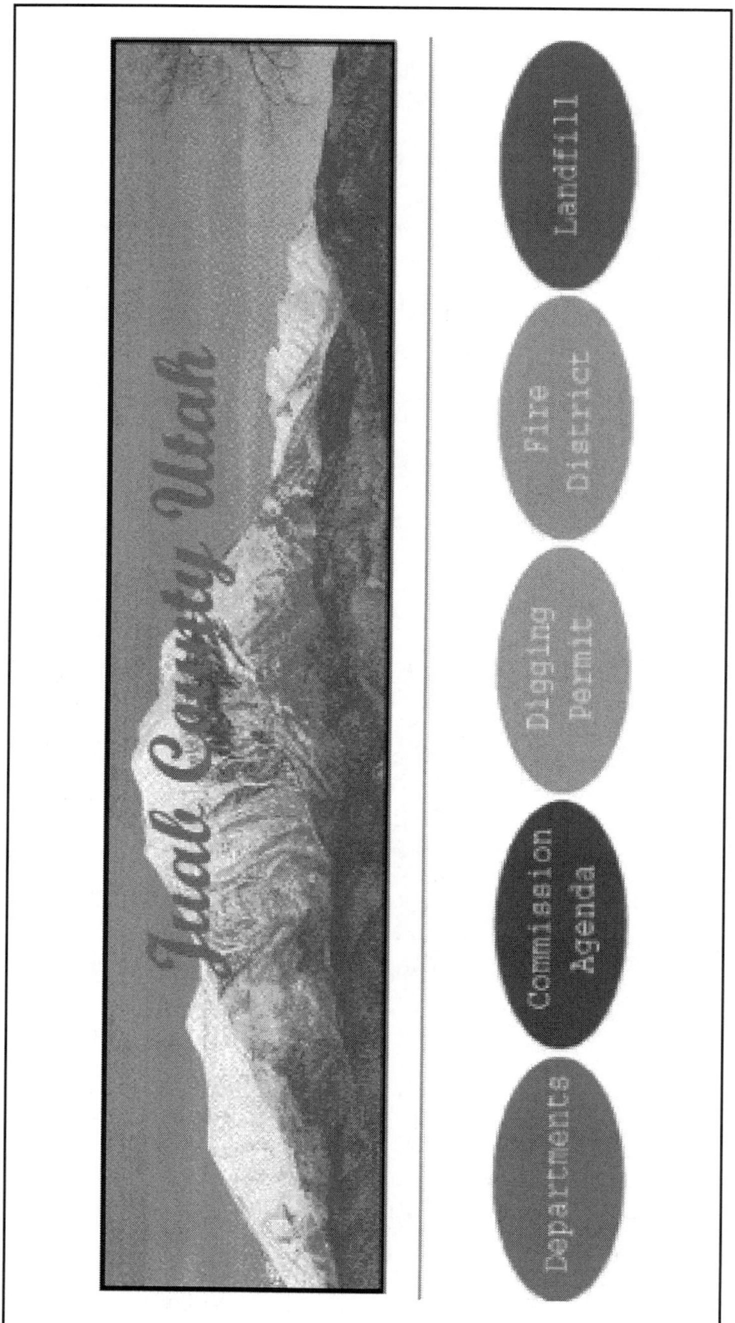

Figure 10. Decoratives used problematically as indicatives.
http://www.co.juab.ut.us

Color can be used indicatively but is problematic because it tends to draw more attention to itself than to what it indicates.

separately indicative button links. Also, the color and typeface choice are in line with the intended purpose of the links, to evoke the agitated feeling typically associated with these forms: viewers are meant to feel agitated and thus provoked to impulsively click the buttons.

Figure 11 gives a rather less suspect example of color successfully integrated with both indicative and informative purposes.

Note that the diagram in Figure 11 is improved somewhat if the colors are somewhat muted, or in other words, less saturated. Thus, the colors can serve a contrastive, indicative function, but at the same time do not create a clash of feeling that disrupts the level-1 unity requirement shared equally by decoratives and diagrams (1-1-1 and 1-1-3).

Image-Indicative Difficulties

Images can be integrated with indicative functions somewhat more easily than purely decorative forms. Unlike purely decorative forms (code 1-**1-1**), images (code 1-**1-2**) have some inherent indicative properties, (as indicated by the "2" in the shorthand code, **boldface** here). The middle "1" in the code means that images are interpreted by means of similarity, but the end result of that interpretation (the final "2") is to point to some physical object. In contrast, more overtly indicative signals (code 1-**2-2**) are interpreted as pointing to physical objects by means of physical adjacency. Consequently, images can be used as indicatives so long as

1. they call less attention to themselves as decoratives than whatever they point to as indicatives and
2. what they point to as images by similarity is in line with what they point to as signals by physical adjacency.

Indicative images can easily misfire unless these conditions are met. As mentioned in the discussion at the end of chapter 2 (Figure 17), if too large and too flamboyantly colored, indicative images can create feelings of clash and generally call undue attention to themselves.

Generally speaking, an indicative image should be smaller, often much smaller than a merely decorative image (Figure 18). So-called "thumbnail" images are commonly and effectively used as indicatives. Directly analogous to the use of muted color for indicative purposes (Figure 11), a smaller image calls less attention to itself but can still signal the presence of a link or a piece of information.

The second primary difficulty is illustrated in Figure 12. It is a common mistake to use images to call attention to adjacent information, in which the content of the image fails to clearly match the content of the adjacent information, or in which the image very poorly represents what will be found if an adjacent link is followed.

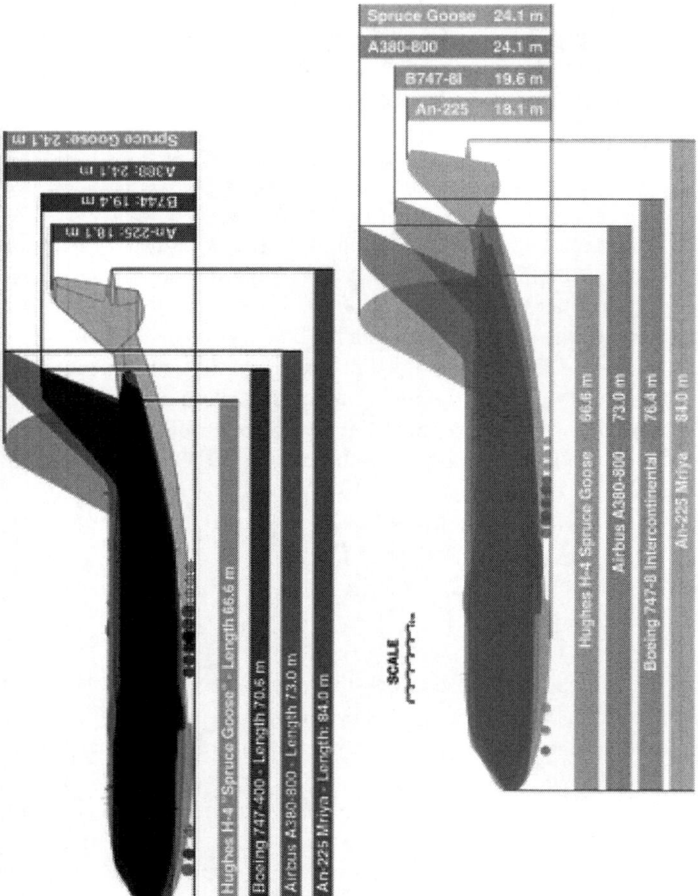

Figure 11. Decoratives integrated with indicatives.
http://commons.wikimedia.org/wiki/Image:Giant_planes_comparison.svg

Muted colors (if limited in number) can effectively support indicative lines.

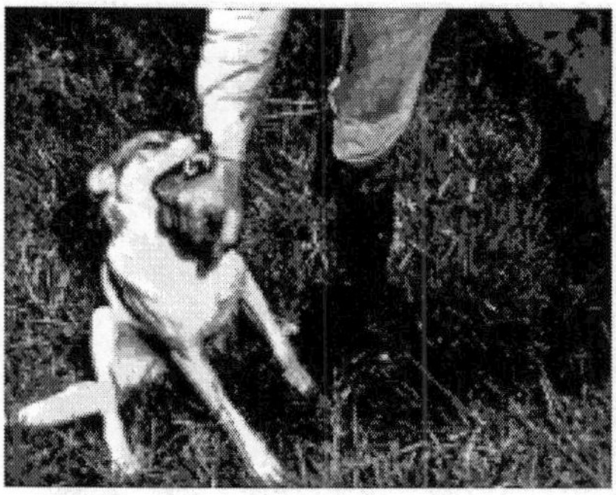

You cannot get rabies by petting an animal. You may get rabies from a scratch if the animal, such as a cat, was licking its paw before it scratched you. (Remember that the rabies virus is found in the saliva of an animal).

In the United States today, about 93 of every 100 reported cases of rabies are in wild animals. Most of the cases are found in raccoons, skunks, bats, and foxes. Raccoons are the most common wild animal with rabies today. Cats had the largest number of reported rabies cases in pets.

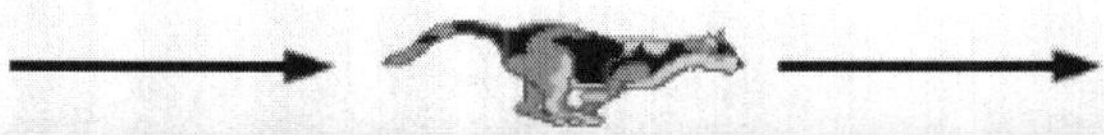

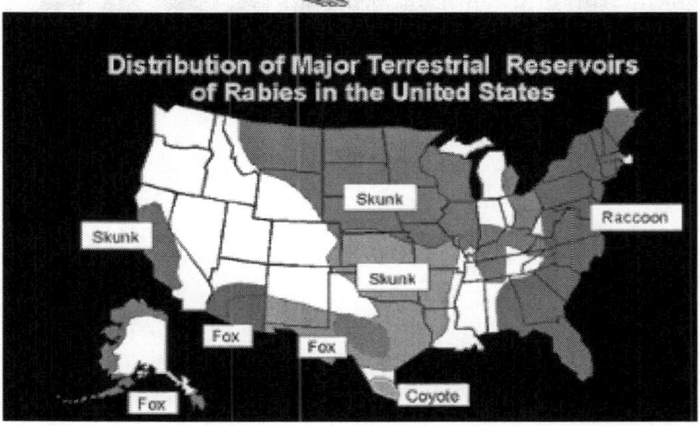

Figure 12. Images that fail to match adjacent information.
www.cdc.gov (Oct. 2006)

It is a common mistake to use images to call attention to adjacent information, in which the content of the image fails to clearly match the content of the adjacent information.

The essential problem illustrated in Figure 12 is one in which the image points one way, by similarity, but points in another, conflicting way by physical adjacency. This violates the formal unity requirement for static signals, and this, in turn, translates into viewer confusion. The rabies explanation (from the Center for Disease Control Web site, cdc.gov) is accompanied by an image of a biting dog, but the adjacent text is discussing how rabies cannot be contracted just by petting a dog (which, at an emotional level, is contradicted by the image). Similarly, the animated image of a running cat separates the verbal explanation for a map from the map itself, showing the distribution of common rabies carriers by region (foxes, skunks, etc.), none of which actually are cats.

As was noted earlier, when an indicative is put in motion, it calls even more attention to itself. If the content of the animation has nothing to do with the information the animation points to, or is adjacent to, it becomes an especially strong, especially obnoxious distraction, forcing viewers to briefly pay attention to something that does not serve their needs.

Bullet Point and Animation Difficulties

Besides the common overuse of images and decorative forms in PowerPoint-style slide presentations, as noted in previous chapters, the PowerPoint program itself is constructed to automatically bullet text and then to easily animate those bullets. Audience members are moved to follow bits of text as they buzz around the screen. However, such features often lack clear indicative goals, some guiding purpose behind the use of such elements, beyond the PowerPoint authors' desire to impress the audience with the mere fact that the elements are present. Unfortunately,

> PowerPoint's basic functions encourage the display of information in short, chunky bulleted lists, and can enable the subordination of content to format by forcing users to reduce complex ideas and data into small, low-resolution chunks that can be seen from the back of a large room. (Clark, 2008, p. 53)

Like the heavy-handed use of color and imagery, these indicative PowerPoint elements can easily become distracting, perhaps irritating, and perhaps even dangerous substitutes for useful information. For example, Blokzijl and Andeweg (2007) found in their PowerPoint study that information in short, bulleted sentences was not necessarily remembered even as well as longer sentences on the slides. They cite other studies (p. 1) suggesting that overbulleted text actually *impedes* cognition, among these studies was the now fairly well-known analysis by Tufte of the engineering communication problems leading up to the Columbia shuttle tragedy (www.edwardtufte.com; see also Doumont 2005). Key information is easily buried in the bullet-heavy slide format (Figure 13a).

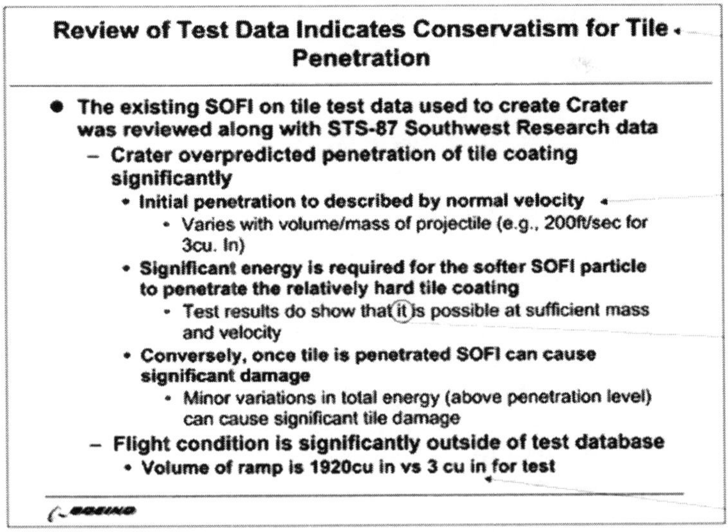

(a) www.edwardtufte.com

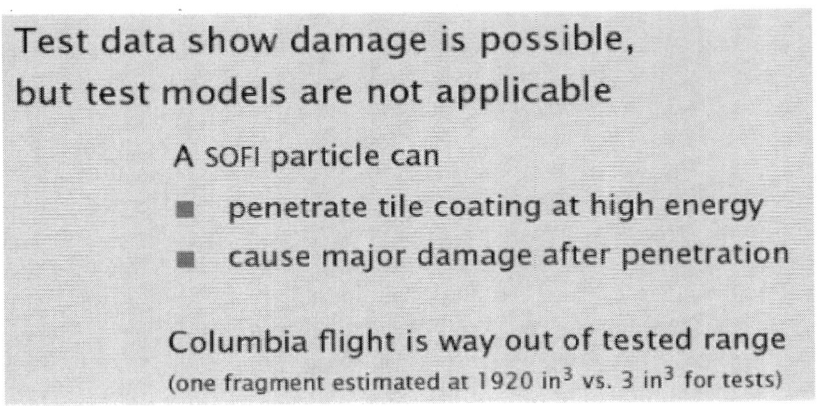

(b) Doumont, J., 2005, p. 65.

Figure 13. Bullet-heavy slide obscures critical information.

Too many hierarchical divisions impose excessive interpretive effort on viewers (a) such that critical information is overlooked. The number of divisions (and the number of words) must be reduced to clarify the message (b).

The conventional wisdom has been that long bullet lists are manageable if broken into hierarchies like those in Figure 13a:

One main point (title)
  One main subpoint with two main sub-subpoints.
    One sub-subpoint with three sub-sub-subpoints
      Two sub-sub-subpoints with parenthetical sub-sub-sub-subpoints.

Our analysis suggests that nested hierarchies do little good and actually amplify the level of effort required to process the slide. The visual requires more than four or five discrete indicative actions (including the action of moving from one hierarchy to the next); viewer fatigue sets in and the key information goes unnoticed (the last two lines of the slide). Therefore, as Blokzijl and Andeweg's findings suggest, a single coherent block of text would be preferable to the very tangible stop-and-go effect created by the slide.

Doumont (2005) suggests, and we concur, that the effective solution to the level-of-effort problem is illustrated in Figure 13a. Contrary to critics such as Tufte, we would maintain that information can be usefully presented in the PowerPoint format, but the visual designer must take responsibility for the key points of information in the slide, and reduce the separate indicative-action steps to a manageable number (i.e., three or four separate points, as in Figure 13a).

In contrast to this simplification solution, it is often supposed that problems created by too many bullets on a slide can be solved with more complexity; for example, with animated effects. If the bullets slide into view, one by one, it seems as though they might be easier to for an audience to process. The problem with animated bullets is that they literally force audience members to shift their eyes to follow the moving bullet. Viewers are forced to see each moving bullet point, but the total energy required of viewers to process the entire sequence of bullets is just that much greater, and fatigue can set in even sooner, such that the audience tunes out altogether and misses the point of the presentation. They remember the moving elements but not what they only opaquely pointed to. The presentation may be perceived as "boring," as even Microsoft admits (Blokzijl & Andeweg, 2007, p. 1), but the underlying truth is that the presentation is exhausting.

Animation can be a useful tool, particularly if the animation directly aligns with information content, such as when a moving bar corresponds with an actually increasing quantity, but the animation must be used with restraint. To force anyone into action without their informed consent (i.e., that the audience knows they are being led somewhere useful) can in many cases be considered a breach of ethics.

To take a more clear-cut example, most Web users have at one time or another come across Web page links that promised to lead to, say, a coupon code for a reputable online store, but instead the link leads to a survey of some kind, or a fee-based program that the user is required to sign up for in order to receive the coupon, or whatever was promised by the initial Web page link. Such scenarios are clearly understood as ethically suspect because the user is misled into a series of futile actions that divert them from personal goals to serve the goals of the Web page link designers. Our point here is that the same kind of ethical problem is created when PowerPoint viewers are led into a morass of animated bullet points that distract from the apparent purpose of the presentation.

CONCLUSION: TOWARD AN INDICATIVE ETHIC

Ethical improvements to indicative communication generally depend on shifting our orientation away from raw feeling → image → action thinking (i.e., the leading edge of the Peircean system, as shown in Figure 8) and toward diagrammatic modes of presentation (as discussed in chapter 3) and indexical modes of presentation (chapter 5). Improvements can be accomplished simply by deploying visual signals such as bullets and animation in selective, measured ways that directly call attention to essential information.

The intended contrasts between separately signaled information points blur together when there are too many bullets or other signaling strategies. Research quite clearly indicates that preferably four and surely no more than five such signals in any single visual should be used (Cowan, 2001); otherwise, problems of memory, separate recognition, and viewer fatigue develop. The situation is made worse if each and every bullet is animated, and sound effects may further distract an audience from grasping the overall point of information. Thus, an epistemological (i.e., truth-seeking) ethic is frustrated by excessive use of visual signaling strategies that overwhelm and fatigue viewers.

The author of such excessively signaled visual information is in effect physically dragging the audience through the information rather than introducing them step-by-step to more manageable bits on the path to each viewer's personal discovery of the truth (or error) in presented information.

Information not presented in an intelligible form can constitute an ethical breach, particularly if the author means to obscure information by drowning viewers in the sheer level of detail, but it is also an ethical problem if the author is simply unwilling to exercise due diligence to separate more essential from less essential information. In this regard, the level of detail presented by long indicative lists should be considered similar to the level of detail presented by images. The typical solution in either case is a reduction in detail, a focus on relevant detail, and a manageable number of clear contrasts. In other words, lists should be typically made more diagram-like and less image-like in terms of total detail presented.

The problem lies in how the bullets are selected and presented, rather than the strategy of using bullets at all, as some have tried to claim: "Tufte argues that the extensive use of PowerPoint by NASA and Boeing contributed to the space shuttle Columbia disaster by reducing dense, critical technical information to over-abbreviated bullet points" (Clark, 2008, p. 52).

Abbreviation is certainly not the problem, in our view, since brief but well-chosen points are better remembered than points that are lost in detailed discussion. We've used bullet points ourselves, several times in this book. A bullet list—even an animated list where bullets fly across a screen and land in place with some plunking sound—might be deployed effectively if the audience really needs to have their attention focused on a small number of specific points, which the audience is able to hold in memory, and which are worth holding in memory.

The problem is not the bullet points *per se* or the animation *per se*. The question is whether signals and action triggers of any kind serve a coherent purpose shared by audience and authors alike. The audience is not likely to embrace a visual-design goal requiring them to actively follow an excessive number of indicative signals. As we will see in the next chapter, however, the problem of purely indicative signals and physical effort can be largely resolved if the indicatives are organized into informative sequences, sequences that allow viewers to more efficiently search and locate specific indicative items in a large informative array.

CHAPTER 5

Informative Indices

In the previous chapter, we discussed the more basic indicatives. These are signals and action triggers such as bulleted lists and Web page links, colored tags, and animation. We emphasized that these indicatives, in their more basic forms, should be used in measured doses—ideally only three or four such elements in any single visual display—to avoid overworking users.

However, to mediate complex information, signals and triggers often must be deployed in greater numbers, but this can only work visually if these action-provoking indicative forms are merged with informative organization and content (Figure 1).

Figure 1 demonstrates the difference between an unorganized, excessively long, and horizontally extended button menu and essentially the same list alphabetized as an *index*. The logical sequence of the index enables users to locate their personal point of interest more quickly, passing over what would otherwise be distracting, separately signaled entries, the otherwise unorganized buttons.

NOTE: Mere alphabetization is *not* sufficient to redeem any overly long random array of elements. The initial word of an alphabetical entry must be a keyword familiar to the user, and the items collected together in the list must constitute reasonably complete inventory. These requirements will also be discussed in this chapter.

We will refer to these informatively configured indicatives as *informative indices*, defined as visual forms that acquire meaning by physical adjacency (the *indicative* part) but also further defined as visuals in which the kind of meaning acquired includes *informative* or, in other words, *propositional* content, propositions generally of the form *point X precedes point Y*.

A well-constructed *informative index* can overcome many of the working limitations of the more basic signals and action triggers (chapter 4). If indicatives are organized by higher informative content, then users can more readily process more than three or four elements or steps in one visual display. Informative organization serves to unify and make coherent what would otherwise be

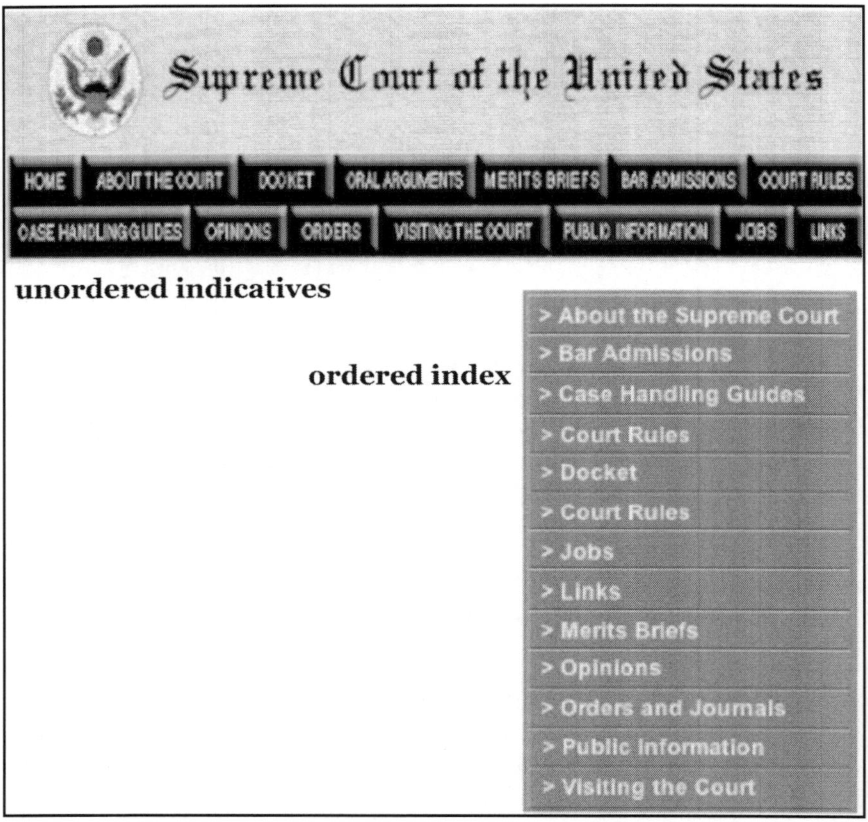

Figure 1. Unordered indicatives vs. informative reference index.
www.supremecourtus.gov

The logical sequence of the index enables users to locate their personal point of interest more quickly, passing over what would otherwise be distracting, separately signaled entries, the otherwise unorganized buttons.

unrelated and separate actions provoked by indicatives. Informative structure makes these actions more manageable in terms of memory load. Informative indices include sequenced or logically ordered labels, table entries, indexical lists, and also visuals that signal users to engage in ritualized action sequences; that is, narrative patterns with *socially coded* (i.e., conventionally recognized and expected) beginnings, middles, and ends.

Narrative, ritual sequences have considerable value as organizers of complex information points. Stories are of course one of the main repositories of cultural knowledge, and stories can be applied at the microlevel of corporate culture as well. A carefully woven narrative can provide a framework superior to mere bullet points for organizing business presentations (McKee & Bronwyn, 2003; Shaw, Brown, & Bromiley, 1998). Ritual sequences typically have an initial trigger, such as the "Once Upon a Time" typical of fairytales. Another such *ritual-sequence* trigger is the shopping-cart motif commonly used to guide users through online purchases (e.g., amazon.com, drugstore.com), or the "start here" button that engages a long sequence of action steps such as the filling out of an online survey.

Such visual motifs engage a narrative pattern or, in other words, a brief story, a story primarily giving a point-after-point understanding, telling viewers how the sequence of stages is to be understood and navigated, for example,

Register to activate cart →
select first item for cart →
select second item for cart →
→ proceed to checkout →
check total →
complete purchase.

Another example of a ritual sequence index is shown in Figure 2, in contrast to merely indicative action trigger, in this case, a piece of animation. Animated visuals quickly become irritating if they only serve to draw attention, but animation that illustrates a procedure in motion, toward a specific goal, can be very helpful.

Note that every story sequenced by problem-solution or conflict-resolution is an informative index of ritual action, from the mininarratives in magazine or TV advertisements to the multivolume epics of literature, whether in print or on film. As noted in the last chapter, more than one or two isolated, animated elements in the form of flashing signals or moving text will induce significant irritation or fatigue in users, but animation organized in a narrative sequence, such as a cartoon or movie clip, is typically eye-catching without being tiresome, as long as the animation is communicating information of interest.

This higher order of organization once again illustrates the power of an informative indexical sequence to make a long series of animated movements manageable in terms of memory load. Just as the basic indicatives were divided into two main types—signals and action triggers—so likewise the more-complex informative indices are divided into two parallel types, *reference indices* and *ritual-sequence indices.*

- Signals (code 1-2-**2**), which otherwise would just be interpreted as pointers, are "elevated" with informative content and therefore result in labels, tables,

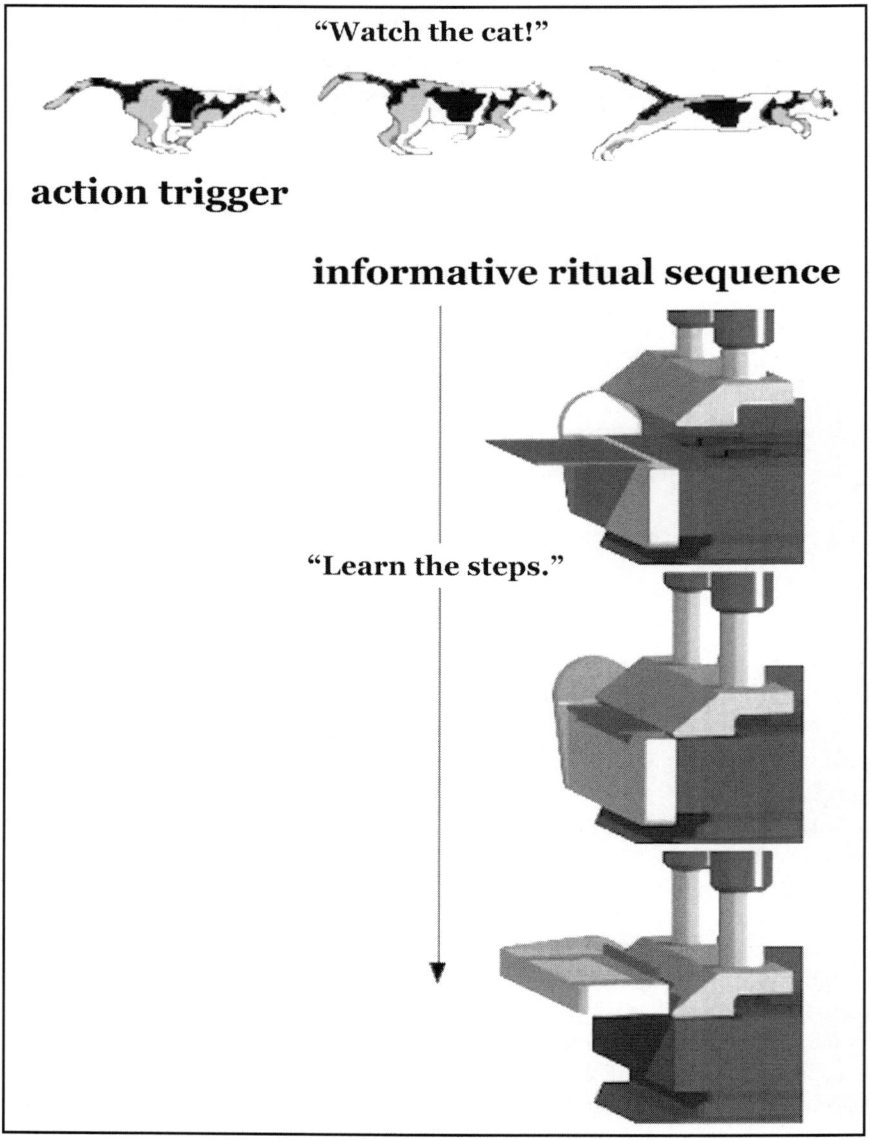

Figure 2. Indicative animation (watching) vs. ritual-sequence animation (learning).

Animated visuals quickly become irritating if they only serve to draw attention, but animation that illustrates a procedure in motion toward a specific goal can be very helpful.

indexical lists, etc., now interpreted as X-before-Y information (code 1-2-**3**). These are Peircean type VI *reference indices*.

- Action triggers (code 2-2-**2**), would otherwise just be interpreted as starting points for raw action (as indicated by the final "2" in shorthand code **boldface** here. However, those action triggers which are "elevated" with informative content and therefore result in narratives, procedures, routines, etc., now interpreted as X-before-Y actions (code 2-2-**3**). These are Peircean type VII *ritual-sequence indices*.

In either case, these informative indices provide organization that enables some degree of complexity for signals and action triggers that otherwise would overwhelm users. This "elevation" of signal to reference index, and action trigger to ritual-sequence index is diagrammed in Figure 3a, and illustrated with examples in Figure 3b. Notice that the relationship between the lower and higher indicatives is exactly parallel, in terms of Peirce's system, to the relationship between images and diagrams, discussed in chapter 3.

As indicated in Figure 3a, diagrams, reference indices, and ritual indices all mediate between direct perception/action and the information contained in words, sentences, and text. As such, reference and ritual indices also have quasi-linguistic properties like those attributed to diagrams in chapter 3, inasmuch as they all

1. include clear contrasts
2. filter out all but relevant details
3. represent generalizations better than specifics

In contrast to images and unordered bullet lists, which do not differentiate the relevant from the irrelevant, both diagrams and informative indices are organized in terms of contrasting points that nevertheless relate to one another in some specifically relevant way. Also, in contrast to images and lists of random specific points, both diagrams and informative indices are directed toward general principles rather than the idiosyncratic properties of individual objects. Even tables of data (a kind of reference index), though they typically display specific points of data, are only useful if they further indicate general trends. So too, ritual sequences are expected to apply consistently to any reenactment of a story, a procedure, or a routine.

We organize and interpret diagrams and informative indices as asserted propositions—*point X is at location Y*, etc.—but we interpret images and lower indicatives as imperative calls to action, such as *look here!* etc. This is the fundamental distinction diagrammed in Figure 3a and illustrated in Figure 3b. However, Figure 3 also reflects what the higher and the lower indicative subtypes have in common, which is that they *actively point*. Put another way, each indicative feature of a visual must *physically connect* to something outside of itself. Recall that the arrow signal (→), like any indicative, represents a call to action, to notice

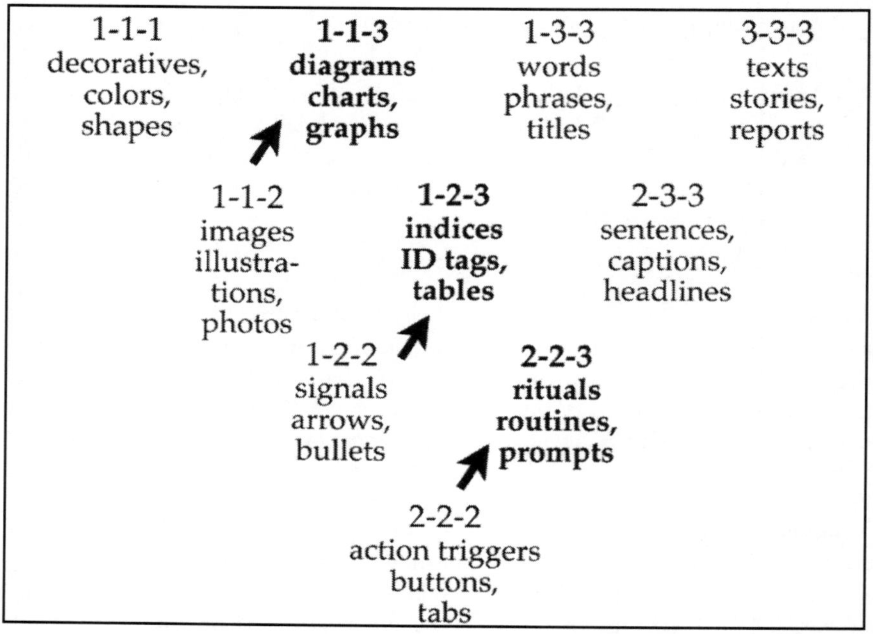

Figure 3a. Decorative/indicative forms elevated to informative forms.

The relationship between the lower and higher indicatives is exactly parallel, in terms of Peirce's system, to the relationship between images and diagrams, discussed in chapter 3.

whatever is physically next to and/or in line with the arrow or other indicator. It is important to also remember that white space itself serves as an effective indicative signal, provoking viewers to notice, for instance, subdivisions within larger sets.

The four kinds of indicatives are →

→reference indices
→social-code rituals
→signals
→action triggers

While the "lower" signals and action triggers point to things wholly outside of themselves (words immediately adjacent to the arrows and adjacent to the white space above, for example), a "higher" informative index points to other elements within a larger informative structure, a frame for propositional information, as in Figure 4.

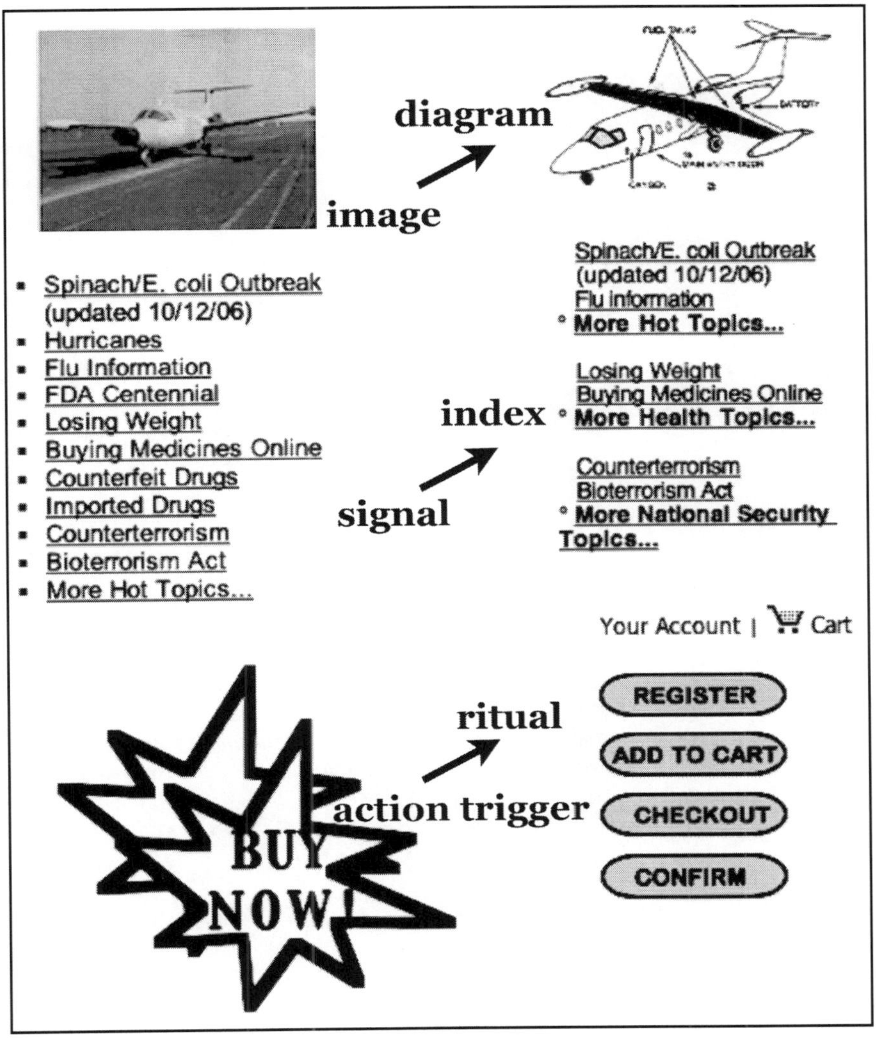

Figure 3b. Illustrative examples, image-to-diagram, signal-to-index, action-to-ritual.

Reference and ritual indices also have quasi-linguistic properties like those attributed to diagrams in chapter 3, inasmuch as they all (a) include clear contrasts, (b) filter out all but relevant details, and (c) represent generalizations better than specifics.

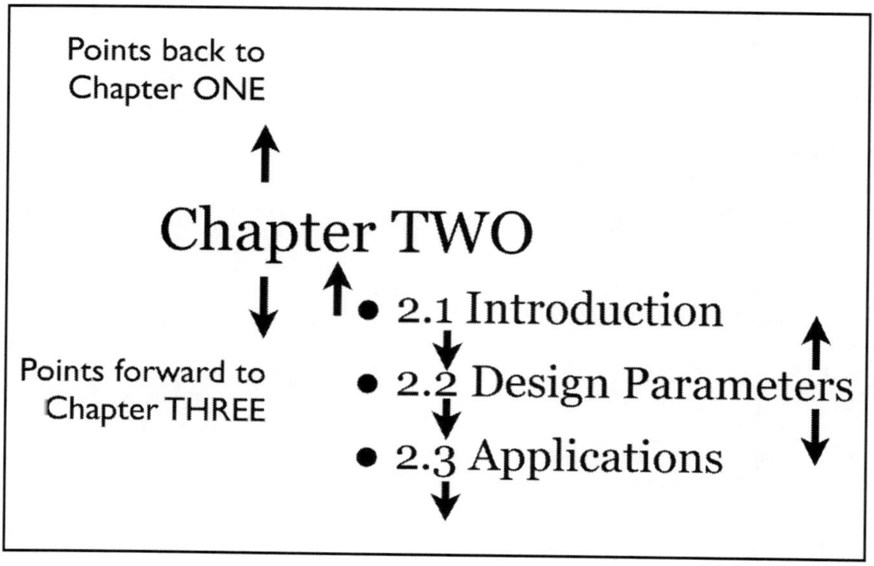

Figure 4. Informative indices pointing to other elements in a larger informative structure.

While the lower signals and action triggers point to things wholly outside of themselves, (e.g., an arrow pointing to an exit door), a higher informative index points to other elements within a larger informative structure.

Note that the larger list of chapters and sections in Figure 4 gives each individual entry a place in a logical sequence—*point Y is AFTER point X*, etc. This logical sequence makes the relatively large number of individual indicatives cognitively and perceptually manageable. In an unorganized list or menu however, so many bullet points or Web page links would be unmanageable in their raw form. A greater number can be managed when embedded within informative structures.

An effective Web page with multiple links and informative points is, in essence, a table or, in other words, a holistically structured informative index. In a well-formed table, the horizontal rows and vertical columns typically have a logical sequence such as first-to-last, highest-to-lowest, A–Z, and the same is true of a well-formed Web page (see Figure 5).

We will now consider in detail the Peircean parameters of these two informative-- index types.

Ranking group	Average R&D funding from industry (%)			Schools with >10.% of R&D funding from industry (number)		
	1993	1998	2004	1993	1998	2004
All universities and college	9.3	8.5	5.9	179	158	101
1–200	7.9	8.0	5.0	50	52	21
1–25	5.6	7.7	5.4	2	4	3
26–50	7.3	6.6	5.0	4	5	3
51–75	6.0	6.7	4.0	3	4	0
76–100	6.5	5.3	3.8	1	2	0
101–125	8.4	7.7	5.0	10	8	3
126–150	8.2	7.8	4.0	9	9	1
151–175	10.8	8.8	5.5	11	9	5
176–200	10.4	13.2	7.1	10	11	6
>200	10.0	8.7	6.3	129	106	80

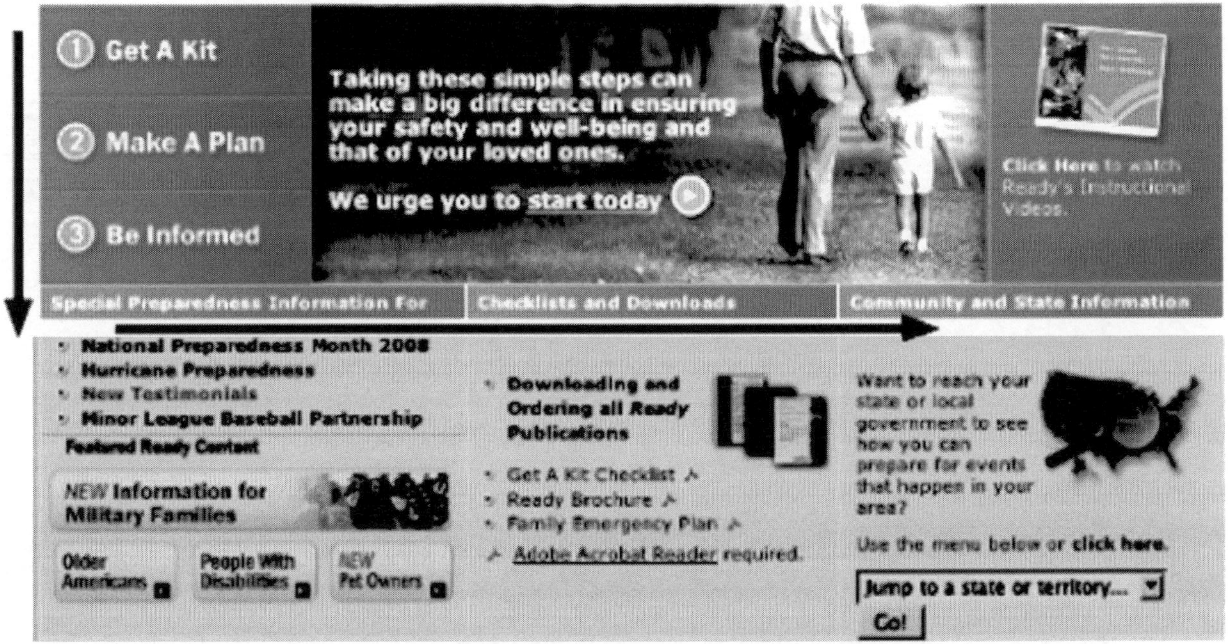

Figure 5. Horizontal/vertical sequencing, the tabular structure in a well-formed Web page.
www.nsf.gov/statistics/inbrief/nsf06328

Usable link menus are either very brief (3–5 items) or are ordered in some way—first-to-last, high-to-low priority, by category, or A–Z. Usable Web pages with complex information must have some kind of table-like structure, with horizontal and vertical categories of information.

INFORMATIVE INDEX DESIGN—PEIRCEAN PARAMETERS

As in the previous chapter, we will consider a related pair of visual sign types simultaneously. Recall that each visual type has three key parameters, based on the Peircean categories:

- its basic form—whether the visual is unified or has distinct phases
- its referential process—how the visual acquires its meaning
- its interpretation type—what kind of meaning the visual form typically acquires

Basic Form

Reference indices (type 1-2-3) and ritual-sequence indices (type 2-2-3) are alike in terms of referential process (physical adjacency) and interpretation type (propositional content), but reference indices (such as the various data points in a table) differ in their basic form from social-ritual indices (such as the various steps in a narrative process).

Like the signals discussed in chapter 4, any reference index is essentially static in form, meaning that the ordered elements of an A–Z index, for instance, or the ordered elements in a table, still form static, unified gestalts. In this way, tables and index lists are not unlike a static diagram showing relationships between parts. The overall organization of the index (or table) should be intelligible as a holistic gestalt. This overall gestalt is what communicates to the user the relative location and thus the meaning of each internal element, its place in the larger whole.

As with all form-level 1 visuals, if the sense of unity is disrupted for the visual, then it fails in terms of basic design. This same requirement holds for reference indices. Schriver (1997) discusses, for example, link lists that fail because content fails to include all reader-expected categories. Shriver, citing Van der Vlist's previous study of Web site usability, notes that readers disliked the list of regions in France because although the Web designer had created links for certain regions, not all regions were listed:

> We can believe that the Web designer left out the other regions [in France] not because he didn't know they existed, but because he had not yet developed their content. But the "say nothing strategy" was rhetorically unwise because it made readers question his knowledge of the subject matter. It would have been better to list all the regions and to point out which links were still under development. (pp. 400–401)

The "still under construction" motif proposed above allows a link list to be incomplete because, instead of a static reference index, the "under construction motif" engages a ritual sequence that is understood to be incomplete at any give step. Like the action triggers discussed in chapter 4, any social-ritual index, in contrast to signals and reference indices, is an essentially nonstatic form, meaning that (usually) only one distinct step of the ritual is perceived at any one instant. Other steps follow in a coded sequence. The overall organization of the socially coded ritual is not a single visual but instead a kind of sequential memory, a diagram that unfolds in time, as it were. Thus, links that aren't working "now" are expected to become active "later."

Of course, rituals can also be transcribed visually as static forms:

A: Knock-knock . . .
B: Who's there?
A: Donya
B: Donya who?
A: Don'ya hate these knock-knock jokes?

The transcript above is a kind of reference index, where we can see at a glance the structure of a typical knock-knock joke—ABABA, etc.—but the joke itself is not fully realized except in performance, or at least in our notion of the performance as a linear sequence. Likewise, a sheet of music is a reference index, but an actual performance of the piece (or a memory of the melody, the note sequence) is a social-ritual index. The table of contents for a novel is a reference index. The scene-by-scene movement of the story (i.e., the plot) is a socially coded, ritual-sequence index, experienced and remembered as a sequence of images in time.

In sum, the formal unity of any reference index contrasts with the formal duality (A → B, B → C, etc.) of any socially coded ritual sequence. Compare Tables 1 and 2.

Referential Process

Apart from these formal differences listed in Table 1, reference indices and social-ritual indices are otherwise alike. The process by which they acquire meaning is invariably by physical connection, as opposed to iconic similarity. Each item in an indexed list is a member of a logical sequence and points forward to the next item and/or back to the previous item in the sequence. Each step in a ritual sequence, like each note in a melody or like each scene in a story, points forward to the next step and/or back to the previous step (Figure 6).

In order to function, the elements of an informative index must be physically located near, next to, or in line with other elements in the overall informative structure.

Table 1. Key Design Parameters of Reference Indices; Shorthand Code = 1-2-3[a]

Basic Form	Referential Process	Interpretation Type
Qualities of unity	Physical adjacency	Propositional relationships
Form Level 1	Reference Level 2	Interpretation Level 3
1	**2**	**3**

[a]Reference indices (code 1-2-3) require formal unity (level-1 form), reference determined by physical adjacency (level-2 reference), and interpretation in terms of propositional information (level-3 interpretation).

Table 2. Key Design Parameters of Ritual-Sequence Indices; Shorthand Code = 2-2-3[a]

Basic Form	Referential Process	Interpretation Type
Dynamic sequence	Physical adjacency	Propositional relationships
Form Level 2	Reference Level 2	Interpretation Level 3
2	**2**	**3**

[a]Ritual-sequence indices (code 2-2-3) require a dynamic sequence (level-2 form), reference determined by physical adjacency (level-2 reference), and interpretation in terms of propositional information (level-3 interpretation).

- In a table or ordered list, the individual elements must be physically next to each other, such that the overall informative structure is apparent.
- In a narrative sequence, each step must closely follow the preceding step in time, such that the flow of the story (i.e., the expected order and organization of events) is apparent.

Without this kind of physical adjacency, the referential coherence of the informative index can easily fall apart. Indexical design choices can therefore misfire if parts of the indexical organization are not visibly connected or if the expected next step of a ritual sequence is interrupted with something unexpected.

Interpretation Type

Interpretation of informative indices passes through the level of (a) feeling and (b) physical reaction, including these levels but further adding (c) informative

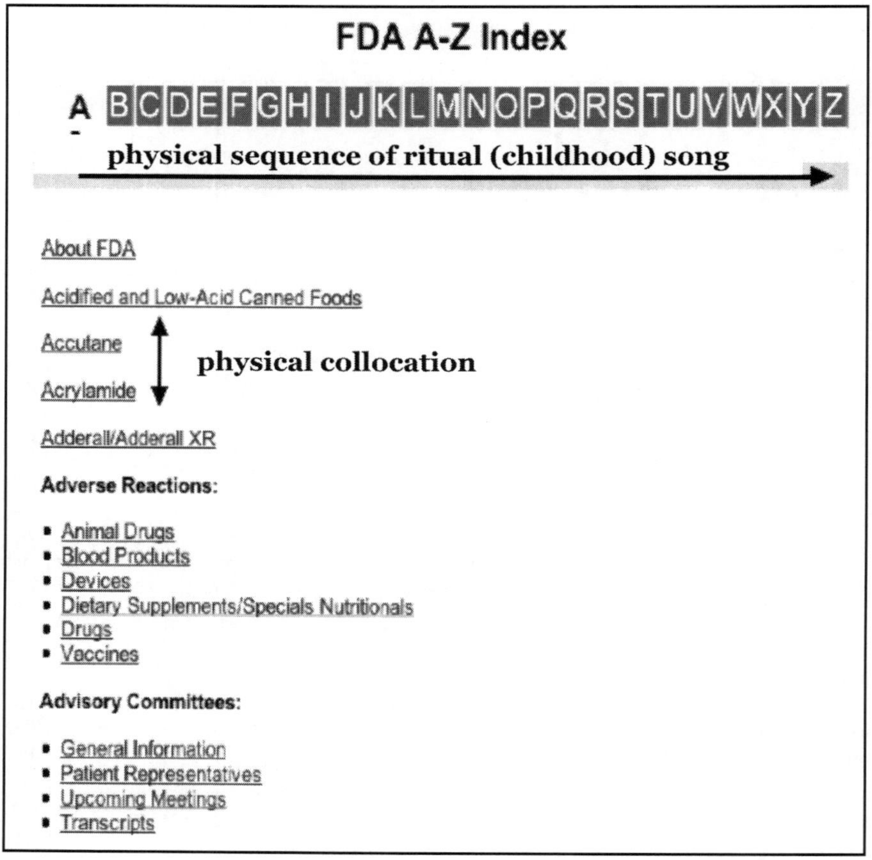

Figure 6. Referential process of informative indices.

Informative indices acquire meaning either through physical collocation (reference index) or physical sequence (socially coded ritual).

relationships. For interpretation to function at this level, the user must be aware (i.e., be informed) about the logical system that organizes the visual. To use an alphabetized index, for example, one must know the alphabet. To use the shopping cart metaphor in online purchasing, one must be at least familiar with the rituals of purchase in actual stores: get a cart, gather items, go to checkout, etc. With that informed awareness, the user, in turn, can extract propositional information from the indexical sequence: There is (or isn't) an entry for *zebra* on the indexical list; one is or isn't required to register at the site before preceding to checkout, and so on.

On this parameter, informative indices radically differ from basic indicatives. Basic signals affect a user by raw physical reaction: the eye is physically and reflexively drawn to a bullet point separated from others by white space, for instance, even more so if the bullet glyph, or the text itself, is highlighted in red, or flashing, etc. Basic action triggers engage actual physical change: a link is clicked and the visual display changes; a different page loads. Those physical effects constitute the interpretation of the basic indicative elements, and they happen regardless of the knowledge state of the user. This is the nature of level-2 interpretation. The user can, at most, choose to ignore or avoid unhelpful signals and action triggers in the visual display.

With informative indexes, level-3 interpretation, some knowledge is required to use them, but the user also has more control, able to deliberately navigate through the reference index or ritual action sequence and able to select the index entry or action step of particular interest.

Recall that the Peircean system shows communication development, starting from the top left corner of the inverted pyramid, but progressing both downward and rightward. Rightward progression from lower points in the system (such as 1-1-2, 1-2-2, or 2-2-2) also involves upward movement. As was noted in the discussion of Figure 3a and 3b diagrams, reference indices, and ritual indices, (1-1-3, 1-2-3, 2-2-3) all in parallel, upward positions relative to images, signals, and action triggers. Referring to the Peircean system in Figure 7, we can summarize the practical implications of this model for information design:

- Just as a diagram is, in a sense, the sum of many different images, a reference index is the sum of many separate indicative signals and a social-ritual index is the sum of many separate action triggers. In terms of memory load, a diagram makes the content of numerous images manageable. Recalling Walsh's (1998) baroque architecture example, a multitude of photographic images is made intelligible by a few line drawings.

  A reference index similarly makes several listed items manageable. A ritual-sequence index, that is, a story, makes several action steps manageable. In sum, visual displays of any complexity must be organized as diagrams, as logically sequenced indexes, and/or as narratively sequenced indexes. Otherwise, usability of the complex visual is compromised.
- Informative indices require physical adjacency between items listed in sequence. If there is too much separation between items in a sequence, the visual loses coherence. In practical terms, this means, for instance, that vertical lists of items are generally preferable to horizontal lists of items. In a vertical list, the first letter and first syllable of each item is physically closer to the first letter and first syllable of the previous and/or the next item in sequence, and users generally identify index items by first letter or syllable, but these are too widely separated in a horizontal menu. Horizontal

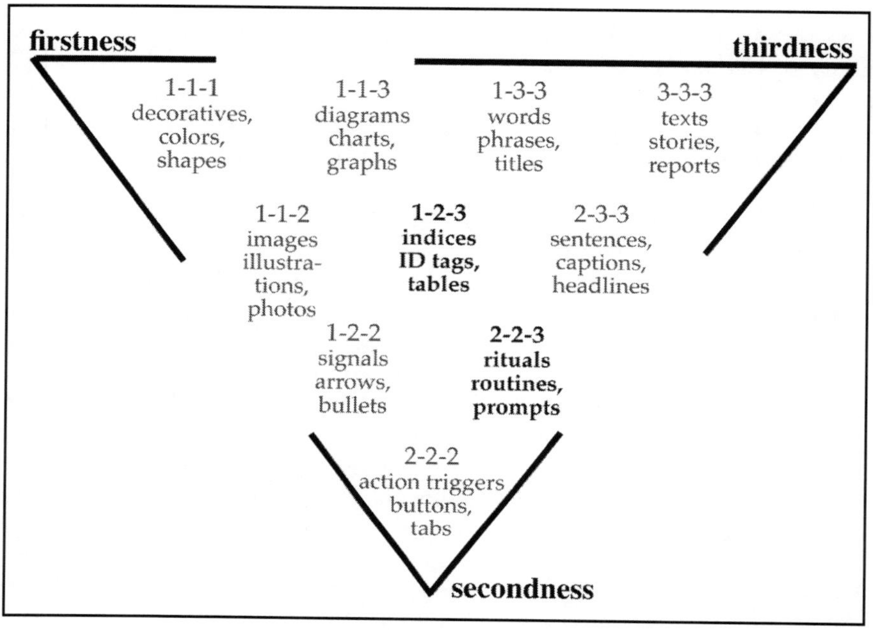

Figure 7. Indices (boldface) located relative to other visual types (muted).

Like diagrams (code 1-1-3), reference indices (1-2-3) and ritual-sequence indices (2-2-3) are intermediate between raw perception-action and language forms.

menus are most usable if there are only a limited number of items, the standing rule for separate signals or action-trigger links. Any indexical sequencing that the horizontal menu might have will be difficult to perceive, at least for first-time users.

- Informative indices require users to already be aware of the conventions governing the logic of the indexical sequence. Sequences like numerical order or alphabetical order or basic narrative conventions (problem-solution or conflict-resolution) will be familiar to most literate users, but these may not provide the best organizing strategy for the information. Other kinds of sequencing can also be used, like general-to-specific, new-to-old, most-popular-to-least popular, lowest-price-to-highest price, and so forth, but it is critical that the visual design indicates to the user what logic is being followed: the sequence that organizes any indexical list or ordered series of action steps. Otherwise, the visual will lose coherence, and usability will be compromised.

We will now consider some examples, illustrating these practical implications in detail.

Indexical Applications and Limitations

To take a common scenario, suppose you are shopping online for a lamp. You come across various Web sites offering several different product lines and styles. From various pages of various shopping sites, you save to your computer the images of candidate choices (Figure 8).

Once you've decided which product you like best, it may prove difficult to relocate it online if all you saved were the raw product images.

The information required to (re)locate the source of an image will consist of, at minimum, one or more reference indices, product number tags, or product description labels that implicitly compare and contrast one item with other items in some kind of ordered matrix, that is, a table or list of items sequenced by specific organizing principles, as shown in Table 3.

In these kinds of situations, the image file itself is most useful if tagged with the Web site of origin, its product number, and its product name. Each of these tags can function as a reference index, useful in locating a particular product. However, note that each product image in Figure 9 is accompanied by reference indices, but that reference information is disrupted by the presentation of multiple items in long, unordered lists. If the reference index tags are noticeably out of sequence, then coherence and usability break down.

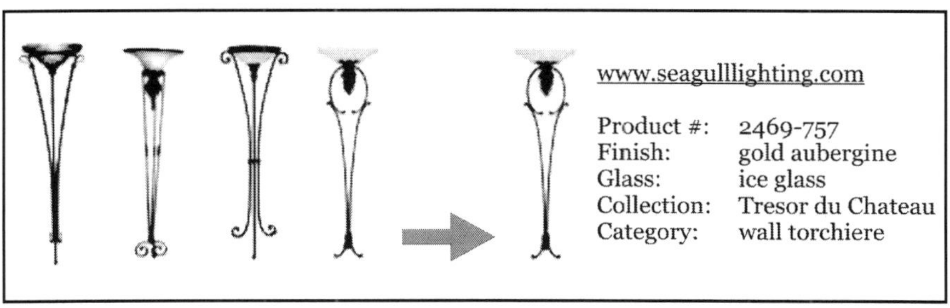

Figure 8. Image vs. Index.

Raw images do not convey information locating those objects; reference indices are required.

Table 3. ID Tags (Figure 8) Converted to Table/Index Entries[a]

Product #	2436-71	2456-85	2469-757
Finish	antique bronze	gold patina	gold aubergine
Glass	None	alabaster	ice glass
Collection	Nouveau	Acanthus	Tresor du Chateau
Category	wall torchiere	wall torchiere	wall torchiere

[a]Reference indices (1-2-3), if represented as identification tags (Figure 8), systematically locate items using specific bases of comparison, comparisons that can invariably be represented in a table or index.

As discussed in chapter 4, any unordered indicative list will act physically on readers, forcing eyes to focus briefly on each listed element. Adding active links to the unordered list only multiplies the workload of users. Users might have to visually inspect dozens or hundreds of list items and actively click on a significant fraction of the links to find what they need.

Among the pieces of bathroom furniture displayed in Figure 9, various ordered comparisons are possible: price, width, number of sinks, or material used (in alphabetical order). However, none of these sequences is used to organize the visual display of inventory here. The separate reference elements with each image are simply part of one link to one product description (i.e., bathroom vanity).

The pages of the Web site are ordered, but this page numbering has absolutely nothing to do with items listed. A user has no way of knowing what type of vanity is presented on any given page and may be forced to search each item and each page to be sure of finding a particular item. An ordered index is always essential to the usability of large blocks of information. The difference between a rummage pile of books and a useful library is only a matter of carefully sequenced reference indexing. To improve usability, any Web site with significant content should, like a library, be reorganized as a reference index so that users need not actively inspect each listed item. A different Web site more effectively designed as a reference index is illustrated in Figure 10.

Another kind of usability problem arises when a page contains certain links or buttons that must be clicked, or clicked in a certain order, to accomplish certain tasks. Difficulty will result when a user overlooks a link or button and thus skips a step in the overall task. For example, as usable as the site exemplified in Figure 10 might be, if users overlook the SORT BY> pull-down menu, then they will have use of only the default listing sequence.

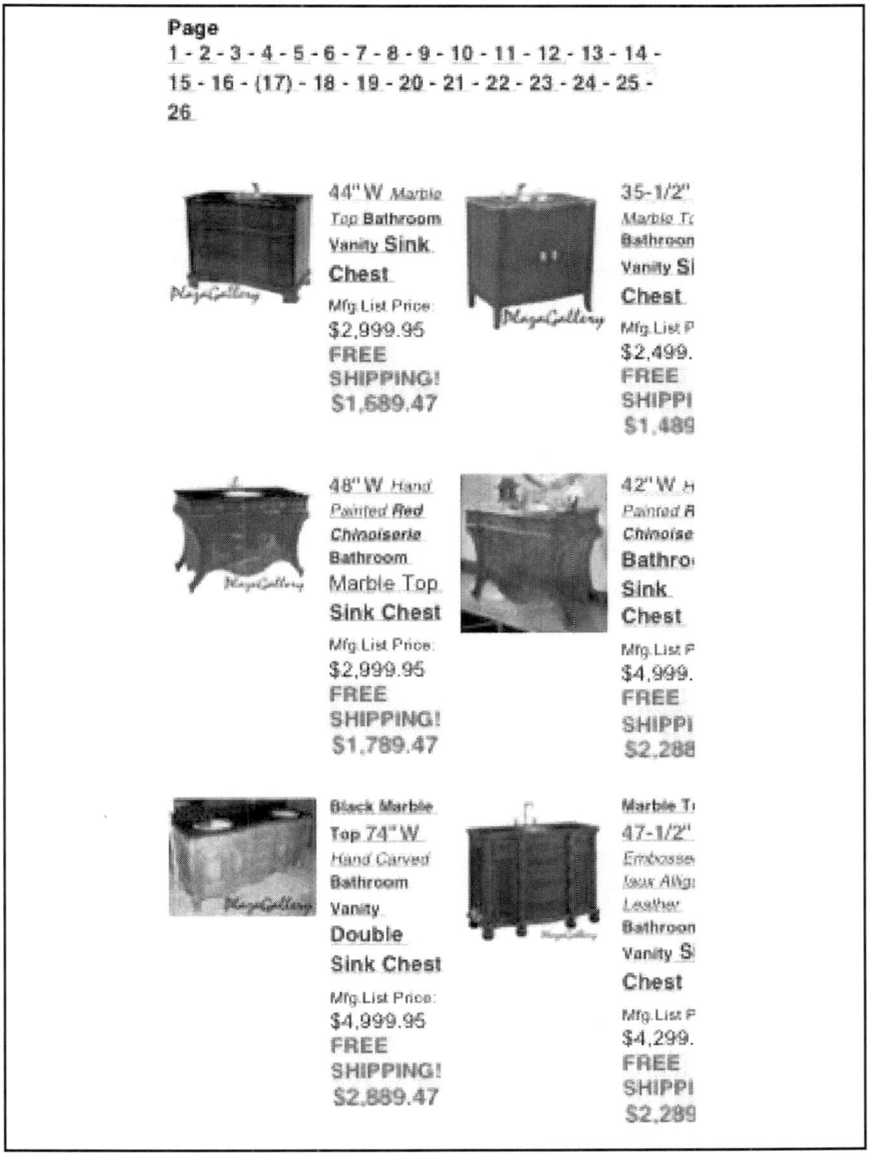

Figure 9. Unusable index entries.
www.plazagallery.com

Each product image is tagged with reference indices, but a lack of logical sequencing disrupts visual coherence.

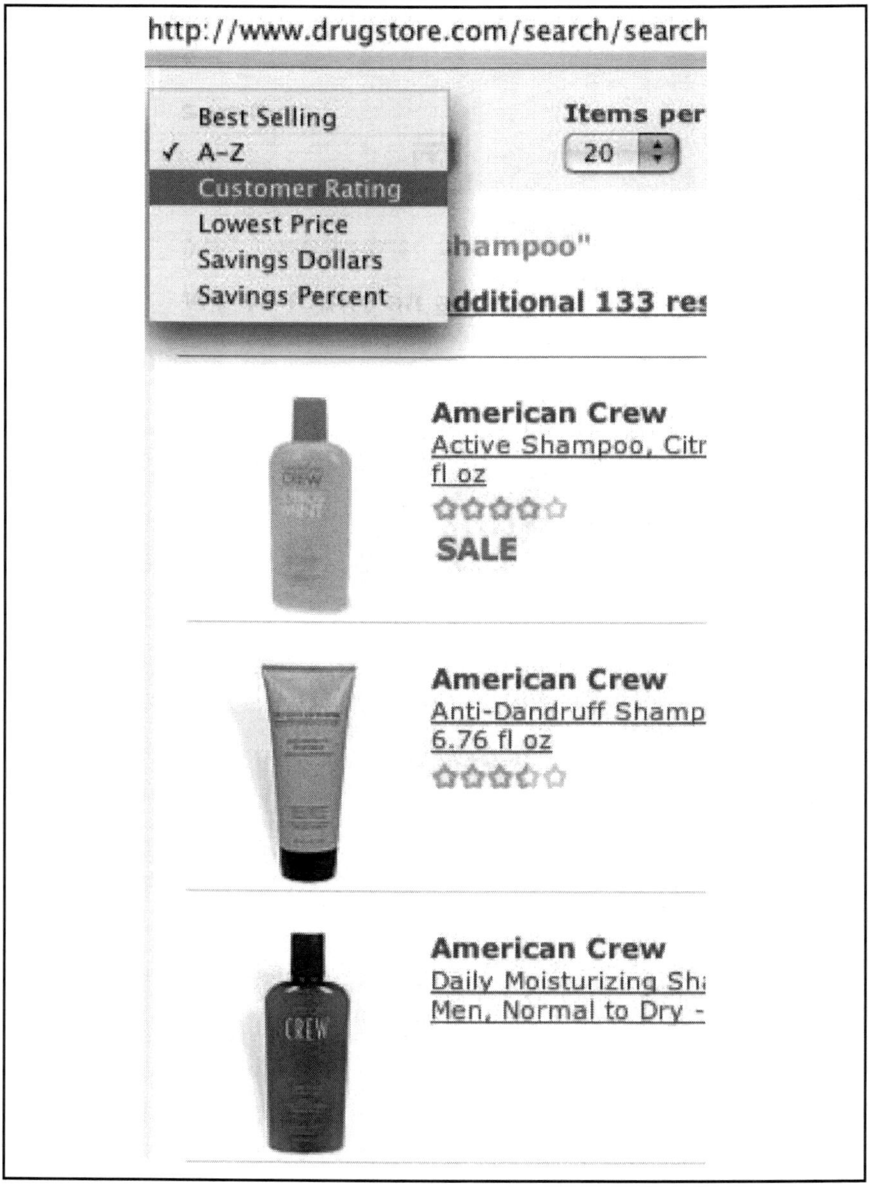

Figure 10. User-adjustable index.

Web site inventory organized as a reference index items: listed in various ordered sequences chosen by users using drop-down menus.

As Kostelnick and Hassett (2003) point out, the usability of any (hypertext) document is contingent to a large degree on users being initiated into and being aware of the implicit codes, customs, and rituals embodied in the text (pp. 3–6). This initiated awareness tells users to look for things like pull-down menus, to expect things like login prompts, to be ready with passwords, to know when it is (or isn't) safe to disclose personal information or credit card numbers, and so on.

All links and buttons that are part of a conventionally determined order are socially coded ritual-sequence indices. Whereas reference indices (1-2-3) organize sequences of visually listed items, otherwise merely signals (1-2-2), social-code indices (2-2-3) organize sequences of action triggers, otherwise merely 2-2-2. Because so much of usability is contingent on established social custom, knowing what button to hit and when because of previously learned social codes, Web designers must either

a. be familiar with current trends in Web design and be careful not to design pages in a manner radically different from action sequences that users have encountered before, or
b. if a radical new design is attempted, it must transparently resemble some ritual pattern that users are familiar with in a separate domain, which they will know to apply to the hypertext as an organizing metaphor.

In other words, innovation in Web design is possible, but if the purpose of the site is action oriented—to buy, sell, instruct, register, etc.—then the site must at least metaphorically resemble those same patterns of activity, and in fact actually sequence those actions as they typically happen to users outside of hypertext. Another good example is the U.S. Postal Service online postage calculator (Figure 11).

Usability problems can emerge if the Web designer introduces anything into the ritual action sequence that interrupts or varies from the familiar real-world ritual and is not visibly tied to some other clearly invoked social-code pattern. For example, many Web sites require shoppers to login to make purchases. It is very common, however, for users to begin shopping, that is, adding items to an electronic shopping cart, before they have actually logged in. Note that this fits the usual shopping ritual as social code. We shop first and then present identification and pay.

A serious usability problem arises when the login command appears at the end of a user's long shopping session, and the login action causes the site to reset, and the user's shopping cart is now empty. This is a good way to lose customers. The point here is that Web site designers cannot afford to simply ignore existing patterns of social action, changing or interrupting them arbitrarily.

If real-world rituals must be altered in visual design, as in an online shopping environment, the designer should make those changes very overt, training the

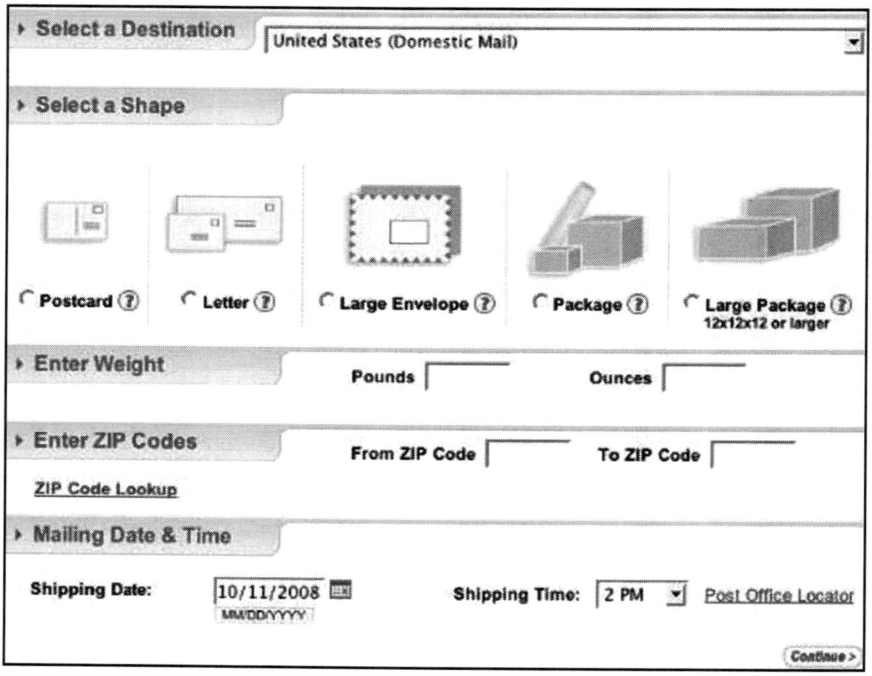

Figure 11. Postal service postage calculation site; good example of familiar process converted into action steps.

Visual display resembling a familiar routing/ritual: sending a package.

user on the spot in the altered ritual, for example, overtly prompt the user to log in before they can begin shopping, thereby invoking other social rituals besides straight shopping—the idea of presenting a membership card (like at a gym or shopping-club warehouse)—before one can enter the member-privileged space. Usability problems are avoided as long as some pattern of social action is invoked that users can recognize, but if no recognizable pattern is invoked by the Web site designer, usability problems multiply.

CONCLUSION: TOWARD AN INDEX ETHIC

Ethical improvements to indexical communication generally depend on making sure that the indexical sequence is transparent, that separate items in the index are in a proper physical/visual relationship to one another, and that the logic of the indexical sequence is apparent and familiar to users.

The current fad of extensive horizontal menus in Web pages goes against this indexical ethic on several points: first, because the horizontal menu follows no logic whatsoever but is instead an unorganized collection of separate signals or action triggers (as, for example, in Figure 1, top). The second problem is that even if the horizontal menu is alphabetized, the critical first letters of each menu item are not physically adjacent to the first letters of other menu items. The third problem is that, given the convention of vertical scrolling and the relative difficulty of horizontal scrolling, the horizontal menu must have a much more limited extent than any vertical menu.

We note that diagrams, reference indices, and ritual indices all share common, quasi-linguistic properties that allow these visual types to mediate between verbal information and direct perception/action. Ethical improvements to informative communication generally depend on shifting our orientation away from an image-driven and unordered bullet-driven orientation in visual design. At the same time, however, complex information must be mediated and visually processed. Thus, we need to consciously move toward these visually mediating modes of presentation, always keeping in mind these essential distinctions:

Decoratives-Images-Bullets-Animation
1. lack of clear contrasts
2. mix relevant and irrelevant details
3. represent specifics better than generalizations

Diagrams-Indices-Ritual-Sequences
1. include clear contrasts
2. filter out all but relevant details
3. represent generalizations better than specifics

Unfortunately, like the distinction between images and diagrams, the critical importance of ordered indices and the power of ritual (i.e., narrative) sequences is not well or widely understood. This lack of understanding can lead to ethical difficulties (as we have already seen in this and in previous chapters) and a disjunction between the author's informative responsibility to an audience—usually best served by diagrammatic and indexical modes—and an author's personal desire to put as little effort as possible into the display of information, which tends to result in the dumping of unorganized lists and links in PowerPoint presentations and Web page designs. Though pervasive, these tendencies can be successfully resisted if we can consistently demonstrate visual design competence through well-constructed diagrammatic and indexical presentations. In the next chapter, we'll see these same considerations actively applied in the configuration of text. In other words, language itself has Peircean visual-design properties.

CHAPTER 6

Words, Sentences, and Text

In previous chapters, we've discussed what are traditionally understood as visuals: decorative forms, images, diagrams, signals, animation, indexical tables, and ritually ordered visual sequences. As noted in the Introduction, it is also our specific agenda in this work to show that words, sentences, and longer texts are also properly understood as visual configurations of information.

Language and visuals have been traditionally treated as separate categories. For the ancient Greeks and Roman rhetoricians, this separate treatment may have seemed like a plausible and adequate distinction, back in the day when the verbal consisted almost exclusively of formal oral speeches and the visual consisted almost exclusively of decorative iconography. For modern information designers, this simplistic distinction is hardly plausible or adequate.

Since the 4th millennium B.C.E., when the ancient Sumerians began transcribing spoken words as marks on clay, and down to the present day, information systems of any sophistication have had to merge the visual with the verbal, whether in print and on video screens. This is because the visual configuration of the verbal is utterly critical for its intelligibility.

As noted in chapter 4, white space between words and even individual letters is a kind of visual indicative. In addition, punctuation marks like periods and commas are indicative signals, properties of visual language. To remove white space and punctuation signals from a written text is to remove intelligibility, particularly if the text is complex in its content.

Just as the right configuration of indicatives is essential to intelligible text, so too, there is a subtle but indispensable decorative (i.e., aesthetic) component. In the Peircean model, we can define more precisely this decorative component, recalling chapter 1, as the feeling associated with the text, based on perceived similarity between textual form (for example, balance, symmetry, contrast, variety) and feelings of playfulness, agitation, or resolution typically felt in conjunction with those forms. In addition, it matters that a text visually resembles, that is, is similar to other texts in users' experience, especially

federalcommunicationscommissionrulessection95paragraph421whomaysign
applicationsforacitizensbandradiolicenseapplicationsamendmentstheretoandrelated
statementsoffactrequiredbythecommissionshallbepersonallysignedbytheapplicantifthea
pplicantisanindividualbyoneofthepartnersiftheapplicantisapartnershipbyanofficerifthea
pplicantisacorporationorbyamemberwhoisanofficeriftheapplicantisanunincorporatedas
sociationapplicationsamendmentsandrelatedstatementsoffactfiledonbehalfofeligiblegov
ernmententitiessuchasstatesandterritoriesoftheunitedstatesandpoliticalsubdivisionsthere
ofthedistrictofcolumbiaandunitsoflocalgovernmentincludingincorporatedmunicipalitie
sshallbesignedbysuchdulyelectedorappointedofficialsasmaybecompetenttodosounderth
elawoftheapplicablejurisdiction

Figure 1. Text with essential visual indicatives (space and punctuation) removed.

Intelligibility suffers (same text as Figures 2 and 3 plain text, no spaces or punctuation).

those that have similar content. If that similarity is disrupted, as for example, when a text is set in an unexpected typeface and format, then aesthetic comfort (i.e., feeling) is disrupted and intelligibility is threatened. It's difficult or impossible to read a text when we are feeling physically ill, confused, upset, or agitated; likewise, there is difficulty if users are made to feel mentally or emotionally uncomfortable by the presentation of the text, as in Figure 2.

Decorative considerations therefore always matter in a text. It does not follow, however, that mere decoration alone will make a text acceptable or usable. This is a typical novice mistake, as we discussed in chapter 1, to give in to the temptation to ornament a text with images, backgrounds, borders, etc. that are largely irrelevant to text content. Such decoration constitutes distraction and also is likely to disrupt the necessary visual similarity between any given text and other texts with similar content in users' experience.

As important as the right indicative and decorative configurations are to the intelligibility of verbal information, we also consider words, phrases, sentences, and longer texts as visuals, and part of the same system of visuals that we have been discussing. This is because language is organized by the same principles used to configure effective diagrams, tables, and ritual-sequence indices. Recall from chapter 3 that effective diagrams fundamentally emphasize contrast. Recall from chapter 5 that effective reference indices and ritual-action indices place each internal element in a logical sequence.

Starting at the most basic level of individual words, contrast and ordered sequence are very much the hallmarks of language: the difference between any two words like *pit* and *bit*, or *pit* and *pin*, etc. is based on very specific sound/letter contrasts. Every word is likewise a specific ordered sequence of sounds/letters. Sentences are similarly built on the contrast between noun and verb, subject and predicate, function words (*the, is, whether*, etc.) vs. content words (*hat, gyrate, impressive,* etc.), and so on. Moreover, sentences are essentially ritually ordered sequences of these contrasting elements: article before noun, subject before predicate, and so on.

Rules of contrast and sequence likewise apply to longer blocks of intelligible text, the fundamental contrast between an expected order of question and answer, topic and development, hypothesis and results, etc. It is precisely these contrasting elements, where each ends and another begins in an expected sequence, that are signaled by the section headings, paragraph breaks, typeface choices, etc. in a well-configured, intelligible text.

It is precisely when we find ourselves unable to track, visually, the essential pieces of a text, as in Figures 1 and 2, that the text loses coherence and stops communicating with us. The deliberate visual configuration of the text, as in Figure 3, helps readers to keep visual track of the essential text elements.

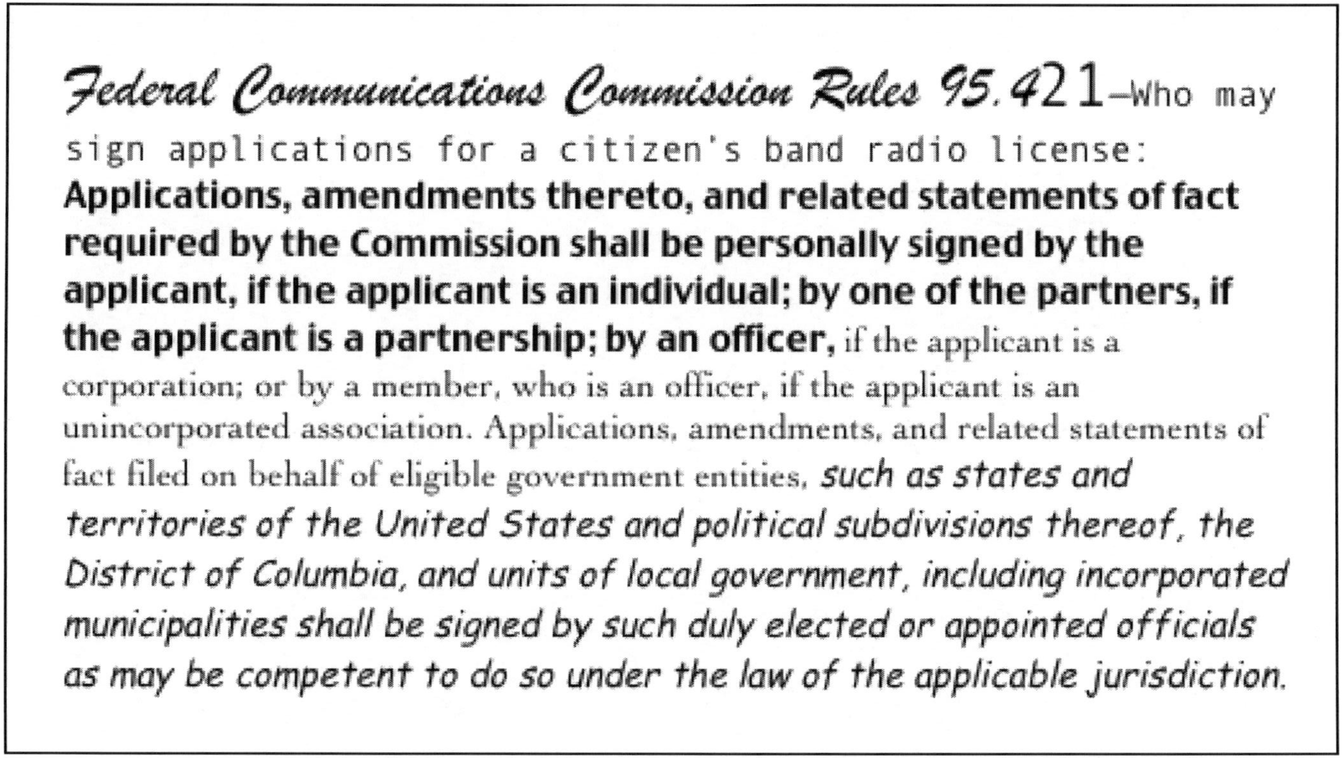

Figure 2. Familiar aesthetics (i.e., decorative properties) removed from example text.

Disordered textual decoratives cause reader discomfort and prevent information extraction.

We will now consider in detail the Peircean parameters of these three primary form-types in language: words, sentences, and extended text. When examined in detail, we find that any of the three language form-types has one or two properties distinct from other visual types; but each language form-type is nevertheless constructed out of the same basic parameters of signification—form, reference, and interpretation—and for each of these parameters there is a first, second, and third level of complexity.

LANGUAGE DESIGN—PEIRCEAN PARAMETERS

In the previous two chapters, we considered visual sign types in pairs, signals and action triggers (chapter 4), and then reference and ritual indices (chapter 5). The remaining Peircean sign types, words, sentences, and text, we will discuss as a set of three. Recall that each of the 10 visual types has three key parameters, based on the Peircean categories:

- its basic form—whether the visual is unified or has distinct phases
- its referential process—how the visual acquires its meaning
- its interpretation type—what kind of meaning the visual form typically acquires.

In turn, each of the parameters has three possible values, one of three levels of complexity, again corresponding with the Peircean categories,

1→2→3:

Basic Form:
static unity > dynamic sequence > *proposition development*

Referential Process:
similarity > physical adjacency > *coded-system relationships*

Interpretation Type:
feeling > physical action > *informative relationships*

Proposition development and *coded-system relationships* are the third levels of complexity in form and referential process for visuals. These third levels of form and reference have not been discussed in previous chapters, and these third levels of complexity are what make language form-types (words, sentences, and text) distinct from other visual types.

However, we have encountered the third level of interpretation, *informative relationships*, in other visuals. Diagrams, reference indices, and social-ritual indices all are interpreted just as language forms are, in terms of informative relationships (i.e., truth-valued propositions), as discussed in chapters 3 and 5.

Federal Communications Commission Rules
Section 95, Paragraph 421

Question: Who may sign applications for a Citizen's-Band (CB) radio license?

Answer:
Applications, amendments thereto, and related statements of fact required by the Commission shall be personally signed by

*the applicant, if the applicant is an individual;
*one of the partners, if the applicant is a partnership
*an officer, if the applicant is a corporation
*a member, who is an officer, if the applicant is an unincorporated association.

Applications, amendments, and related statements of fact filed on behalf of eligible government entities, such as states and territories of the United States and political subdivisions thereof, the District of Columbia, and units of local government, including incorporated municipalities shall be signed by such duly elected or appointed officials as may be competent to do so under the law of the applicable jurisdiction.

OR better still...

FCC Rules— sec. 95 para. 425

How do I sign my CB license application?

(a) If you are an individual, you must sign your own application personally.

(b) If the applicant is not an individual, the signature on an application must be made as follows:

· For a partnership -- one of the partners must sign;
· For a Corporation -- one of the officers must sign;
· For an Association -- one of the members who is an officer must sign;
· For a Governmental Unit -- an appropriate elected or appointed official must sign.

Figure 3. Text configured to optimize intelligibility.

Text is made more intelligible with visible proposition-development structure: headings, a few parallel bullet points, and paragraph breaks.

For each parameter (form, reference, and interpretation), if we represent these parameters horizontally and their levels of complexity as numbered values, we have our standard representation of the Peircean parameter settings. Tables 1, 2, and 3 indicate the settings for words, sentences, and text.

Basic Form

As indicated in Tables 1, 2, and 3, words, sentences, and text (the language-form types) are all alike in terms of referential process and interpretation type. These three modes of language differ only in terms of basic form.

1-3-3: The basic form rule for individual words (or multiword phrases that are not sentences) is exactly the same as the rule for decorative forms, for images, for diagrams, for signals, and for reference indices. That is, any well-formed word (or nonsentence phrase) is organized around *qualities of unity*; the word or phrase presents itself visually as a single, static gestalt, just as decoratives, images, diagrams, etc. do (Riley & Parker, 1998).

Table 1. Key Design Parameters of *Words* (or Nonassertional Phrases); Shorthand Code 1-3-3[a]

Basic Form	Referential Process	Interpretation Type
Qualities of unity	Coded system	Propositional relationships
Form Level 1	Reference Level 3	Interpretation Level 3
1	**3**	**3**

[a]Words and nonsentence phrases (code 1-2-3) require formal unity (level-1 form), reference determined by coded semantic systems (level-3 reference), and interpretation in terms of propositional information (level-3 interpretation).

Table 2. Key Design Parameters of *Sentences* (or Propositional Assertions); Shorthand Code 2-3-3[a]

Basic Form	Referential Process	Interpretation Type
Dynamic sequence	Coded system	Propositional relationships
Form Level 2	Reference Level 3	Interpretation Level 3
2	**3**	**3**

[a]Sentences or, in other words, assertions (code 2-3-3) require a dynamic sequence (level-2 form), reference determined by coded semantic systems (level-3 reference), and interpretation in terms of propositional information (level-3 interpretation).

Table 3. Key Design Parameters of Coherent, Multisentence Texts; Shorthand Code 3-3-3[a]

Basic Form	Referential Process	Interpretation Type
Propositional development	Coded system	Propositional relationships
Form Level 3	Reference Level 3	Interpretation Level 3
3	**3**	**3**

[a]Coherent, multisentence texts (code 3-3-3) require visible proposition development (level-3 form), reference determined by coded semantic systems (level-3 reference), and interpretation in terms of propositional information (level-3 interpretation).

2-3-3: The basic form rule for sentences, in contrast, is exactly the same as the rule for action triggers and for social-ritual sequences. That is, a sentence must consist of at least two distinct phases. Traditional grammar captures this basic form rule by saying that a sentence must have, at minimum, a noun and a verb. More precisely, we would say that every sentence by definition has a subject element that gives way dynamically to a predicate element in exactly the same way that clicking a Web link dynamically loads a separate Web page or the same way that the first line of the knock-knock ritual gives way dynamically to the response:

Knock-knock ==> *Who's There?*

Subject: *the cat . . .* ==> *What about the cat?*
Predicate: . . . *needs to be fed.*

In other words, if you walk up to someone and just articulate a subject phrase out of the blue, like *the cat*, she or he will likely respond with *what cat?* or *what about the cat?* seeking out the predicate, which is automatically called for by the subject. This is directly analogous to the automatic response sequence initiated by a knock-knock joke.

3-3-3: The basic form rule for extended texts does not apply to any of the other visual types. Even so, the text-form rule of *proposition development* is very much a visual rule. To see any group of two, three, or a thousand sentences as one text, a language user has to visualize mentally, and/or actually see on the printed page, a development structure in which the truth of one proposition is supported by (at minimum) two other propositions. This is illustrated by the premise-premise-conclusion structure of a classical syllogism or, in other terms, the *data-warrant-claim* structure of Toulmin-style argument:

This bean is from Bag A. —premise 1 or *data*
All the beans from Bag A are white. —premise 2 or *warrant*
Therefore: This bean is white. —conclusion or *claim*

Any two-line argument—one premise and one conclusion—invariably has a third, implicit proposition, an unstated premise. In classical terms, this is of course an *enthymeme*:

Ramos's car will cost $5,000 to repair. —premise
Ramos's should buy a new car. —conclusion

Ramos's car isn't worth $5,000 ← implied premise

On the other hand, two stated premises often imply a third statement, which is the conclusion.

Ramos's car will cost $5,000 to repair. —premise
Ramos's car isn't worth $5,000. —premise

Ramos should buy a new car. ← implied conclusion

Not all texts are premise/conclusion or data/warrant/claim arguments in the typical sense of *argument*, but Peirce used the term *argument* (in a very abstract sense) to designate what is meant here by proposition development: the truth of a proposition being supported (i.e., *developed*) by at least two other propositions. This basic form rule applies to all coherent units of language larger than a single sentence. When a text consists of only two visible sentences, overtly asserting only two propositions, there is invariably a third proposition implied, or else the two sentences will not be perceived as a coherent whole text. To borrow the example of a minimum plot from E. M. Forster,

The king died.
The queen died. —a mere sequence of events, an uninteresting story.

The king died.
The queen died *of grief.* —a minimum, coherent plot.

In Forester's minimum, coherent plot (1927, pp. 82–83), there is a third proposition implied that unifies the otherwise noncoherent statements:

The king died. —narrative setting
The queen was grieving because the king died. —narrative conflict
The queen died. —narrative resolution

Looked at in most basic terms, the conflict in any plot (expressed as a proposition *X encounters Y*) is developed by information about the setting, the resolution, and any other plot development in between. In another sense, the truth of the implied proposition that *the queen was lonely because the king died* is here supported (i.e., *developed*) by two stated propositions: that *the king died* and that (subsequently) *the queen died.* We can explain (i.e., make coherent and thus make whole) the sequence of deaths by concluding that the queen was lonely. In this sense, we can say that the minimum narrative also has argument structure. Whether a text is narrative (setting-conflict-resolution) or expository (data-claim-warrant), and whether the text is two sentences or a thousand, we expect some kind of three-part form that must be visualized mentally, but preferably is also made visible with headings and white space on the actual page or screen, as for example in Figure 3.

Reference Process

Words, sentences, and texts all acquire meaning through *coded-system relationships*. Like the form rule of *proposition development*, the referential process for words, sentences, and texts is nevertheless a visual process. This language-reference process differs from the processes by which meaning is attached to other visual types, but the reference process for language is in fact identical to the final interpretation state of diagrams, reference indices, and ritual sequences, as X-before-Y kinds of propositions, as we explain below.

Decoratives, images, and diagrams (i.e., iconic visuals) all acquire meaning by apparent similarity (level-1 reference). Signals, action triggers, reference indices, and ritual sequences (i.e., indexical visuals) all acquire meaning by apparent physical adjacency (level-2 reference). The key insight here is that language may acquire meaning through a process that goes beyond mere similarity and physical adjacency, a process that is nevertheless a visual process, that is, a process that requires visual ordering, mapping, and sequencing of words in order for textual content to have coherent meaning overall.

Coded-system relationships have already been discussed as a visual mode in chapter 5. In most basic terms, any information visualized as a table, ordered list, or ritual sequence (such as the plot of a narrative) is a coded system. Recall that the *referential process* for reference indices (e.g., a table of contents) and ritual-sequence indices (e.g., the process of shopping) is physical adjacency. That is, in order for elements of an index or ritual sequence to have meaning, they must physically occur next to each other or one after another. However, the meaning acquired (i.e., the *interpretation* parameter) of reference indices and rituals is not a physical thing or action but rather an abstraction, in each case a kind of code, in each case a system of propositional relationships—*A comes before B, B comes before C*, etc. In a typical alphabetized index, for example, entries beginning with the letter A signify that they should be followed by entries beginning with the letter B, or the letter C if there are no B entries, and so on.

When we move from these indexical forms to language forms, the code-system *interpretation* of these indexical forms (1-2-**3** and 2-2-**3**) becomes in turn the *referential process* by which words, sentences, and text (1-**3**-3, 2-**3**-3, and 3-**3**-3) acquire meaning, as shown in Figure 4.

To determine the meaning of a word (or a sentence or a text), we have to know its place in a coded system. In effect, we are locating each word in a table (i.e., reference index) of intersecting semantic values. Within this matrix of values, meaning for each word (or sentence type or text type) is acquired not by similarity to or physical adjacency to objects, but by its coded-system relationship to other words (or sentence types or text types), as illustrated in Figure 4. These coded-system relationships are readily represented and (in this theory) best understood as implicit visual structures, diagrams within tables, in the mind of the language user (cf. Manning, 1989).

Story plots were discussed as ritual sequences (Peirce type 2-2-**3**) in chapter 5, and so they are, especially when a story is told as a sequence of images without any particular need for words. In the case of a story that is purely made of words, the *interpretation* of story plot as a ritual indexical form (each step in the plot pointing to the next step) becomes, in turn, the reference process and the basic form rule (**3-3**-3) to organize the actual language, words, sentences, and chapters, of narrative text.

Interpretation Type

Like diagrams, reference indices, and ritual sequences, the interpretation of language passes through the level of (a) feeling and (b) physical reaction, including these levels but further adding (c) informative relationships. As with images or other visual signals, so too with words and sentences, we may be equally able to evoke strong emotional responses and even strong physical responses, but language and the other informative visuals are especially suited to convey propositional information as well, claims that can be taken as true or false.

So, for example, a photograph showing the aftermath of an auto collision does not assert that there was or wasn't a collision. The image portrays only the *perception* of damage. The image could have been faked, or the damage to the vehicles could have been created instead by a sledge hammer, etc. It is the photo caption, in words, that asserts the causes of perceived damage, the owners of the vehicles, time and date of the accident, etc.

For interpretation to function at this asserted-proposition level, the user must be aware (i.e., be informed) about the logical system that organizes the language. To use a language, the user must know the language, but this requirement is no different in principle from the kind of prior knowledge needed to interpret a diagram, understand the logical sequence of a reference index, or know what comes next in the sequence of any social ritual.

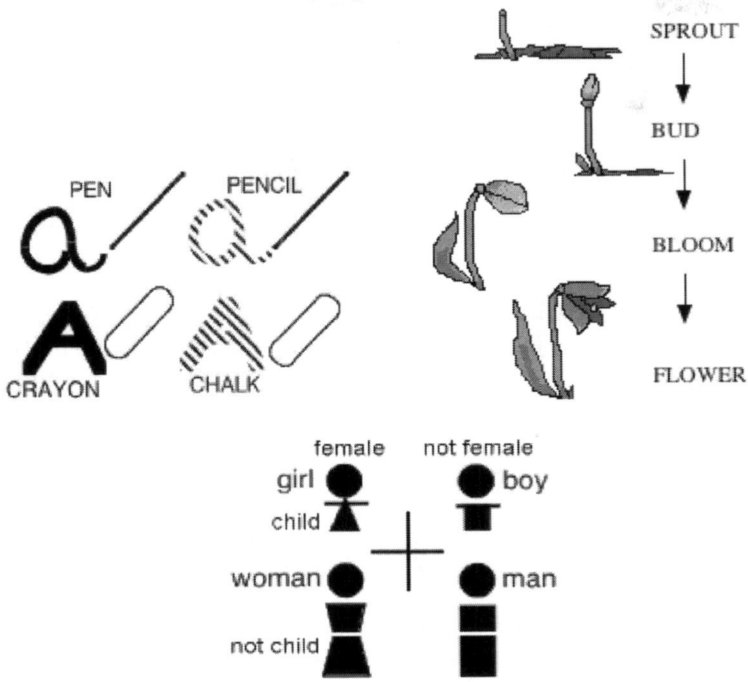

	subject+predicate	single phrase
response expected	**QUESTION** *Is it?*	**COMMAND** *Do it!*
response optional	**STATEMENT** *It is.*	**EXCLAMATION** *Darn it!*

	image-oriented	diagram-oriented
action sequence	**NARRATIVE**	**TECHNICAL PROCEDURE**
thesis supported	**MARKETING**	**SCIENTIFIC/ACADEMIC**

Figure 4. Level-3 reference process for language derived from level-3 interpretation of diagrams, reference indices, and ritual sequences.

Coded-system relationships, i.e., the kind of information presented in a reference index (i.e., a table), enables the referential process by which words, sentences, and text acquire meaning.

To use an alphabetized index, for example, one must know the alphabet. To use the shopping cart metaphor in online purchasing, one must be at least familiar with the rituals of purchase in actual stores: get a cart, gather items, go to checkout, etc. With the same kind of informed awareness, a language user extracts propositional information from words, sentences, and texts. On this parameter, language is not different from other informative visual types.

All visuals with level-3 interpretation require prior knowledge to be used, but the user also has more control of how the visual is used, control in selecting elements of particular interest. Effectively structured language functions just as effectively structured diagrams, tables, indexed lists, and ritual sequences do: to unify and make complex information more manageable.

With the Peircean parameters of words, sentences, and text defined, and their relationship to the definitive parameters of other visuals understood, we are now in a position to explain, by appeal to specific principles and parameters, the difference between effective and ineffective visual language, between effective and ineffective integration of other kinds of visuals with language.

VISUAL LANGUAGE: IMPLICATIONS, APPLICATIONS, AND LIMITATIONS

Recall that the Peircean system shows communication development, starting from the top left corner of the inverted pyramid (decoratives at 1-1-1), and progressing both downward and rightward. Rightward progression from lower points in the system (such as 1-1-2, 1-2-2, or 2-2-2) also involves upward movement. As shown in Figure 5, we find diagrams, reference indices, and ritual indices, (1-1-3, 1-2-3, 2-2-3) all in parallel, upward positions relative to images, signals, and action triggers, but we find words, sentences, and text in upward positions relative to diagrams, reference indices, and ritual indices.

Referring to the Peircean system in Figures 5 and 6, we can summarize the practical implications of this model for visual language design.

The verbal is mediated through the visual: Effectively written text should incorporate effective visual elements as well. Peirce's primary insight here is that language, either spoken or written, is not conceptually processed until and unless it is translated into other visual types, as shown in Figure 6. Information in the form of language is translated, at some cognitive level, into diagrammatic, tabular, and ritual-sequence forms of information before that information can be understood in terms of perception (what we can expect to see, hear, feel, etc.) or in terms of action (how we should plan to act, based on information).

As has been noted in previous chapters, overused decorative and indicative visuals will overwhelm and fatigue viewers. This is the perception-action cascade discussed in chapter 4—the left edge of the Peircean pyramid from 1-1-1 to 2-2-2—and visual presentations that rely too heavily on this edge tend to ethically

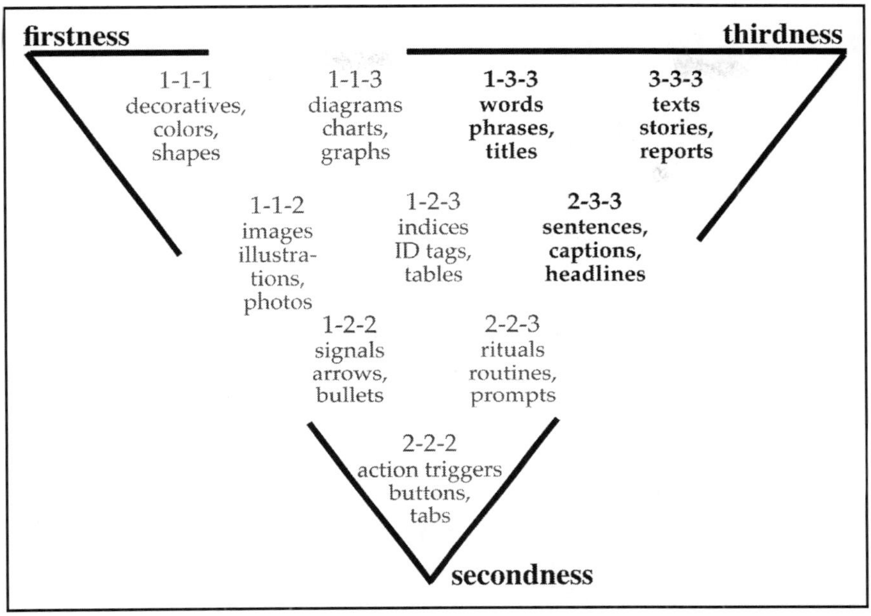

Figure 5. Language forms, i.e., words, sentences, and text (boldface), located relative to other visual types (muted).

Language forms represent the extremes of Peircean Thirdness relative to other visual types, but words (code 1-3-3), sentences (code 2-3-3), and text (code 3-3-3) are still within the same larger system of visual types.

abuse viewers by dragging them through perception-action chains that do not serve both viewers and authors equally.

On the other hand, raw, dense blocks of informative verbal text are difficult or impossible to process. This is the top right corner of the Peircean pyramid, words (1-3-3), sentences (2-3-3), and especially text (3-3-3). Solutions to the various informational difficulties presented by these opposite visual modes (perception-action vs. information) are generally to be found in the middle, in diagrams and indices that can mediate between language and perception-action.

Unity is achieved through parallelism: Individual words and nonsentence phrases (Peircean type **1**-3-3) have the same basic form requirement as decoratives and images (**1**-1-1 and **1**-1-2), and therefore must be chosen with the same considerations of feeling and semantic weight with which we choose typefaces, borders, colors, forms, etc. This translates, among other things, into

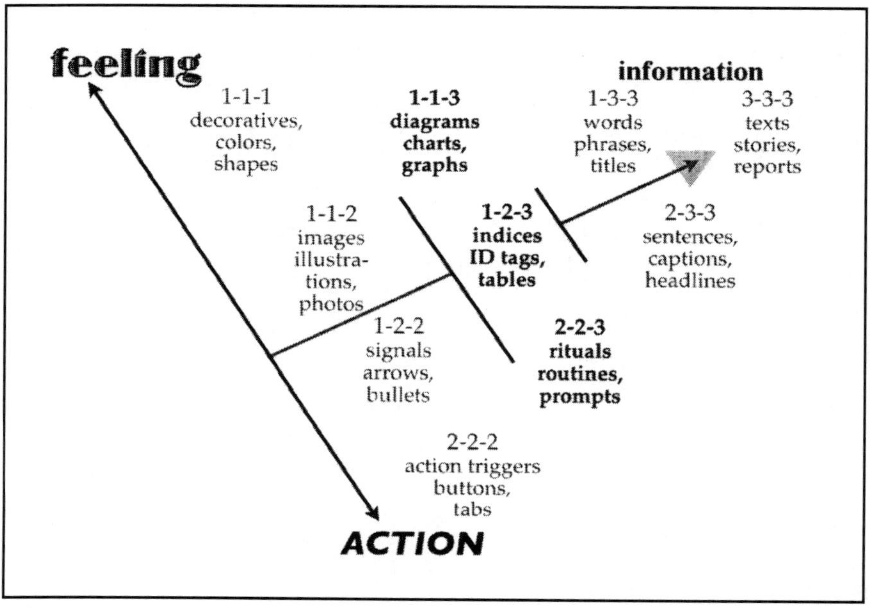

Figure 6. Verbal mediation through the visual.

Perception/action face of the Peircean pyramid (decoratives, images, signals, action) is only connected to language, and vice versa, through the mediation of diagrams, reference indices, and ritual sequencing visuals.

the preference for parallel grammatical phrasing when words and phrases have parallel meaning in a text (Figure 7).

As shown in Figure 7, parallel phrasing reinforces aesthetic unity at a decorative level, which in turn is the same kind of formal unity required of bulleted lists as signals (**1**-2-2), diagrams and tables of information (**1**-1-3) and (**1**-2-3). In each of these visual types, parallelism is necessary for formal unity.

Dynamic sequence is achieved with short, identifiable sentence subjects: Individual clauses and sentences (Peircean type **2**-3-3) likewise have, or should have, significant indicative and dynamic-sequence properties. An audience should be able to connect sentence subjects with objects or ideas in their actual experience. Otherwise, communication fails. Furthermore, the subject creates the expectation of a following predicate, and this predicate should not be long delayed by an overly long subject phrase (Figure 8).

Non-parallel structure

· If you are in a partnership, one of the partners must sign;
· One of the officers of a corporation must sign;
· For an Association, one of the members who is an officer must sign;
· An appropriate elected or appointed governmental-unit official must sign if the governmental unit is applying for a license.

Similar to—

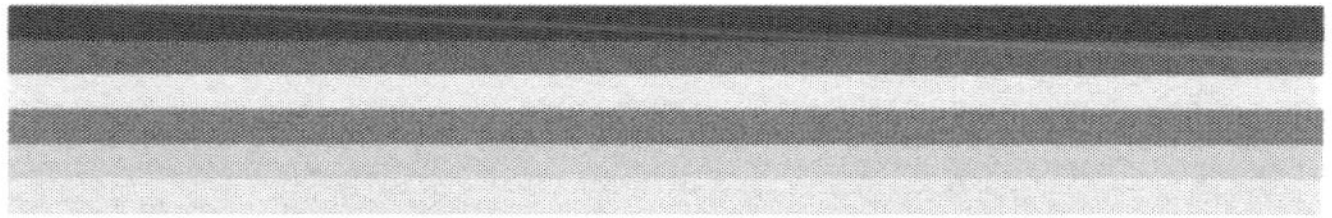

Parallel structure

· For a partnership -- one of the partners must sign;
· For a Corporation -- one of the officers must sign;
· For an Association -- one of the members who is an officer must sign;
· For a Governmental Unit -- an appropriate elected or appointed official must sign.

Similar to—

Figure 7. Phrasal parallelism as a strategy of visual unity.

Lack of parallelism creates clash feeling comparable to clash in a purely decorative graphic (some examples of nonparallel text and parallel revision, compared with some ugly, nonparallel graphic).

< Excessively long subject phrase >

< Applications, amendments, and related statements of fact filed on behalf of eligible government entities, such as states and territories of the United States and political subdivisions thereof, the District of Columbia, and units of local government, including incorporated municipalities>

shall (assertion of sentence)
be signed by such duly elected or appointed officials... (predicate of sentence)

Similar to—

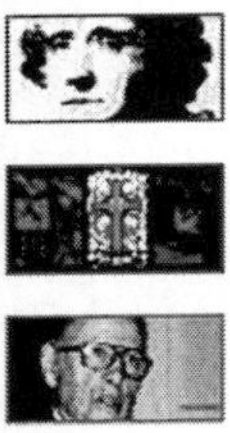

< NON-ADJACENT SIGNAL >

Library Experience at myloc.gov
NOW OPEN! New exhibitions and technologies inspire and engage

Wise Guide to loc.gov
Hurricane lessons, Hispanic heroes, storytelling rugs and yummy candy

Webcasts from the Library
Herman Wouk receives Library fiction prize

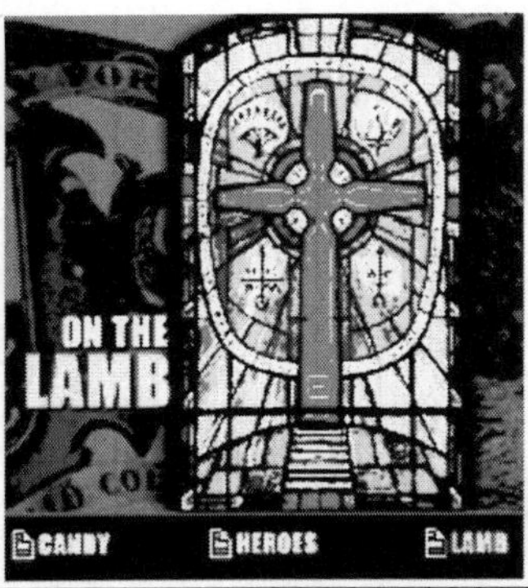

< EXCESSIVELY LARGE SIGNAL >

Wise Guide to loc.gov
Hurricane lessons, Hispanic heroes, storytelling rugs and yummy candy

Figure 8. Sentence intelligibility hindered by overly long subjects; visually equivalent to a violation of physical adjacency in an indicative visual.

Subject-heavy text is comparable to any visual design in which signal elements (in this case, white space) call excessive attention to themselves.

In visual terms, when an overly long sentence subject delays access to a sentence predicate, a formally similar violation occurs—a disruption of the physical-adjacency rule of any signal or action trigger.

Proposition development is visualized in ritually sequenced steps: Whether a text consists of two sentences or thousands, any well-formed text (Peircean type **3**-3-3) develops one proposition with at least two propositions, and those supporting propositions can have further development of their own. In practical terms, this means that text, whether short or long, is invariably divisible into at least three parts:

1. a background of noncontroversial information,
2. a question or issue that provokes doubt and drives the reading of the text forward, and
3. information that addresses the question and supports an answer.

In narrative, this three-part sequence is

1. setting,
2. conflict, and
3. story development and resolution.

In Toulmin-style argument, this three-part sequence is

1. data,
2. debatable claim, and
3. warrant/interpretation of data.

In procedural instructions, this three part sequence is

1. tools and materials needed,
2. task(s) or problem(s) to solve, and
3. procedural steps to the solution.

In an empirical research article (Introduction-Method-Results-Analysis), this three part sequence is

1. the previous research on a topic that has left some question unanswered,
2. competing possible answers to that question, and
3. the answer to that question as determined by the current study, its methods, results, and analysis.

Title: Decorated Visual

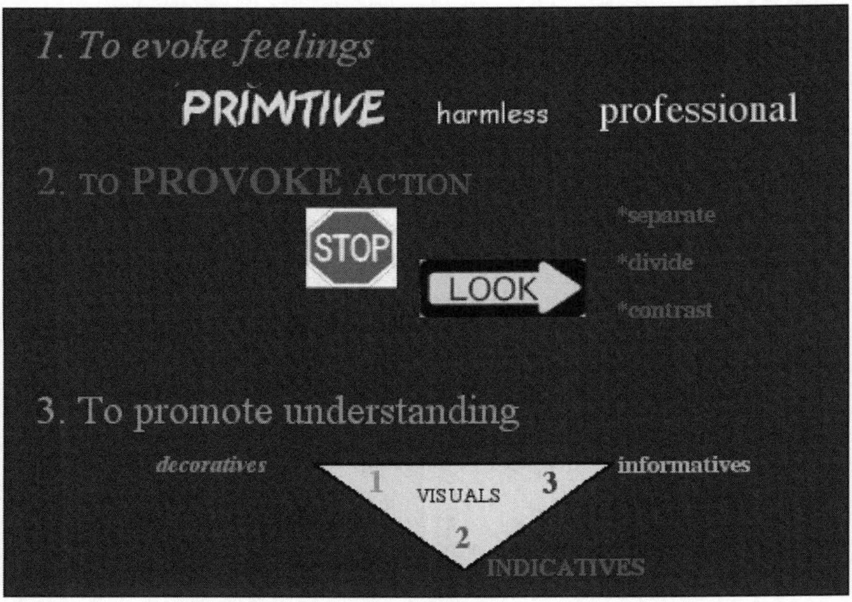

Sentence Caption: Decorative elements, if overused, can interfere with the understanding of a visual.

Figure 9. Relative clarity of complete sentence captions.

Point of a visual is generally clearer with a full-sentence caption than a non-sentence label. In Peircean terms, the diagram, table of data, indicative image, etc. is established as a complete text with proposition development.

Note that the three parts of textual development can occur in different orders, or with 1-, 2-, and 3-type information combined in complex ways, depending on the genre of the text.

As noted earlier (discussion of Figure 3), readers' navigation through these natural divisions of information should in some way be marked out visually in paragraph breaks, sections breaks, heading labels, and so on. In addition, it is important to understand that, while the central proposition of a text probably will be most clear if stated verbally, that is, as a complete sentence, the development propositions of text can be established with diagrams, reference indices, or ritual-action indices, that is, any visuals with propositional interpretation.

Consider, for example, a presentation slide that consists of a diagram, a table of information, or an animated sequence. That single slide can count as a whole text, in the Peircean sense, if the slide is clearly presented as development of a single proposition. This is why complete-sentence captions (clearly asserting one proposition) generally work better than mere phrasal labels (Figure 9).

Thus, recent research also indicates that complete-sentence statements are more effective in PowerPoint-style presentations that mere phrase labels, which do not clearly assert propositions (Alley & Neeley, 2005).

CONCLUSION

Toward a Visual Language Ethic

Ethical improvements to verbal communication generally depend on making sure that the key formal elements of sentences and longer texts can be easily identified visually. This process begins in obvious ways at the word level but should continue just as consciously in the visual separation of distinct textual elements.

- Letter-graphic elements, visually distinct, make up words.
- Words, separated units of white space, make up phrases.
- Phrases, in separate dynamic phases of subject and predicate, make up sentences.
- Sentences, separated by punctuation, make up paragraphs.
- Paragraphs should likewise be separated by indentation and/or white space to make up longer blocks of text.
- Longer blocks of text, broken up by headings and subheadings, should be visibly distinct as well.

Recalling discussion in chapters 4 and 5, note that this six-item bullet list is acceptable because it follows a logical sequence, that is, a ritual index.

Failure to put spaces between words is not different in principle from failure to properly configure text in terms of known ritual sequences, with paragraph breaks and section headings to mark key boundaries in proposition development. They both constitute lapses in the visual design of text, with ethical implications. Likewise, there are ethical issues involved in the failure to provide appropriate diagrams, indices, and ritual sequencing, in order to make complex blocks of verbal information intelligible. When a person hires themselves out as an editor, a Web designer, or an information specialist of any kind, there is an implied promise of competence, and implied responsibility placed upon the information specialist to know what constitutes effective visual/textual design.

Our hope is that, with this book, we have enabled visual/textual designers to be more ethical in their professional responsibilities in using goal-driven strategies of visual information design for the benefit clients and other audiences.

CHAPTER 7

Toward a Universal Terminology and Grammar of Visual Types

SUMMARY

Whether decorative, indicative, or informative in purpose, visuals need to be, above all else, clear. To construct a language for such visuals, to aid in the discussion of visual-clarity strategies, we have chosen the semiotic theory of a philosopher, C. S. Peirce (Burk, 1958; Hartshone & Weiss, 1935), whose 10-part system of sign-icon-index is applicable to 10 core visual types, including written text.

These 10 types can be more broadly divided into three categories of function for visuals: decoratives, indicatives, and informatives. It's worth noting that parts of speech can be grouped by similar Peircean functions (e.g., adjectives/adverbs as modifiers of feeling; nouns as image creators; articles as indicative pointers; and verbs, prepositions, and conjunctions as information connectors, etc.). Likewise, the 10 types of visuals can be grouped by the three simple functions of invoking feeling (decoratives), pointing (indicatives), and providing knowledge (informatives). The 10 types of visuals are, in effect, the essential parts of visual "speech."

In recent articles, we have discussed the application of Peirce's visual design typology in some detail (Amare & Manning, 2006, 2007, 2008; Manning & Amare, 2005, 2006). This book elaborates on this work and discusses more particularly both the need for and the means for developing a consistent terminology of visuals based on their different forms and their diverse purposes. We would suggest that a consistent visual-communication terminology would enable the professional and academic discussions that inform the choice of one kind of graphic over another and further enable the fine-tuning of form in the graphic itself. We further suggest that a coherent interpretation and application of Peirce's 10 sign types is the key to providing a language for visuals and for a better understanding of visual information design in terms of "signs." "A sign," Peirce tells us,

> is something which stands to somebody for something in some respect or capacity. It addresses somebody, that is, creates in the mind of that person an equivalent sign, or perhaps a more developed sign. That sign which it creates I call the *interpretant* of the first sign. The sign stands for something, its *object.* (CP 2.228)

In his work, Peirce proposes several typologies of signs. In fact, there are at least 76 distinct definitions of sign in Peirce's writings, but his 10 sign *types* (CP 2.254–264) form a system that we think is most relevant to visual information design. The 10 types are, in Peirce's original terminology:

I. Rhematic Iconic Qualisign,
II. Rhematic Iconic Sinsign,
III. Rhematic Indexical Sinsign,
IV. Dicent Indexical Sinsign,

V. Rhematic Iconic Legisign,
VI. Rhematic Indexical Legisign,
VII. Dicent Indexical Legisign,

VIII. Rhematic Symbol Legisign,
IX. Dicent Symbolic Legisign, and finally
X. Argument Symbolic Legisign.

Although the terminology above is not transparent, it is generated by a precise grammar, as we noted in the introduction. Even in this book-length treatment, we can discuss only a few of the implications of this grammar.

We have translated Peirce's system into modern terms and applied it to modern-day visuals:

1-1-1. decorative icons (= I, borders, typeface shapes, color, etc.)
1-1-2. image icons (= II, photographs, realistic illustrations)
1-2-2. signaling indices (= III, bullet points, arrows, etc.)
2-2-2. action indices (= IV, Web links, buttons, page tabs, etc.)

1-1-3. informative icons (= V, diagrams, charts, graphs)
1-2-3. reference indices (= VI, tables, figure labels, etc.)
2-2-3. ritual indices (= VII, routines, step-by-step procedures, etc.)

1-3-3. word-symbols (= VIII)
2-3-3. sentence-symbols (= IX)
3-3-3. whole-text-symbols (= X)

Table 1 shows the exact translation between Peirce's terminology into the coded shorthand, which Peirce also briefly laid out (CP 8.376). Peirce did not develop his numerical shorthand any further in his known writings, but its mathematical properties have been reconstructed and discussed at length by Marty (1990) and Robertson (1994).

Moving from column to column, left to right, the level number may rise or stay the same, not jump to a nonadjacent, lower-number row. The shorthand code we've developed in previous chapters is recapped in Table 1. We use that same code to characterize ill-formed visual design. Types *2-1-1, *3-2-1, etc. are ungrammatical and ill-formed. Thus, there are 10 visual types that are well-formed, but there are 17 other possible types that are not well-formed and are, in other words, ungrammatical. For example, we classify as *3-2-1, and therefore *ungrammatical,* a PowerPoint presentation that seems to develop and support a proposition (level-3 form), but only, in fact, drags viewers through several screens of indicative but unintelligible bullet points (level-2 reference process) and leaves the audience with vague feelings, positive or negative, but no actual conclusion as the interpretation (level-1 interpretation).

As theoretically satisfying as the full form of Peirce's system might be, we can also provide a useful simplification for students and basic practitioners of visual design. We can view

- the two "low" icons (I, II) as *decorative* forms and images,
- the two "low" indices (III, IV) as *indicative* signals and actions, and
- the six "high" types, one icon (V) and two indices (VI–VII) and the three symbols (VII, IX, X) as *informatives*.

Table 1. System that Generates Peirce's 10 Classes of Sign (I through X)[a]

Complexity of **Sign Form**	Connection to **Object-Referent**	Complexity of **Interpretation**
		1. Qualisign
	1. Iconic	2. Sinsign
1. Rheme	2. Indexical	3. Legisign
2. Dicent	3. Symbolic	
3. Argument		

[a]Types are defined by one value from each column; each column value has a level number (1, 2, or 3). I = 1-1-1, II = 1-1-2, III = 1-2-2, IV = 2-2-2, V = 1-1-3, VI = 1-2-3, VII = 2-2-3, VIII = 1-3-3, IX = 2-3-3, X = 3-3-3

The situation portrayed in Figure 1, though helpful as an introduction, is somewhat oversimplified because most visuals will contain decorative, indicative, and informative features to some degree.

In this simplified view (Figure 1), we have to maintain that visuals that are *primarily* decorative will not successfully indicate, and visuals that are *primarily* indicative will not successfully inform. However, what varies from visual to visual is the relative degree of each feature. In Peirce's 10-type system, the exact proportions can be specified, but such precision is not always necessary. Decorative-heavy or image-heavy visuals may still indicate and inform, but not strongly and not effectively, in most cases. Informative visuals, like diagrams or tables, conversely, may have minor decorative features (having to do with preserving gestalt unity), but these decorative aspects must not predominate or overwhelm the informative function.

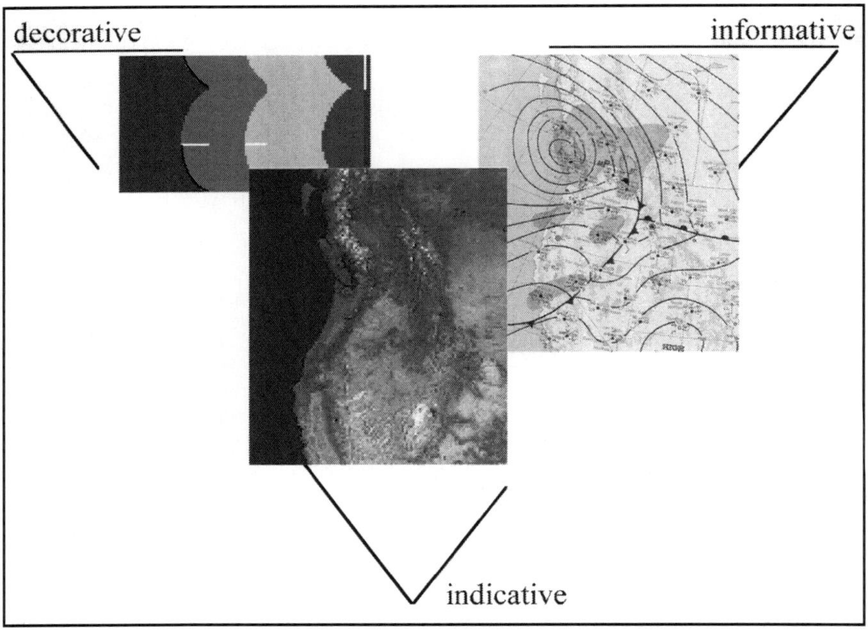

Figure 1. Simplification of Peirce's 10 Types: decoratives, indicatives, and informatives.

For many practitioners, it may be sufficient to impressionistically weigh relative degrees of decorative, indicative, and informative effects, rather than determine precise levels of firstness, secondness, and thirdness in Peirce's 10-class system.

It should be especially noted that the words, sentences, and longer texts of verbal language (i.e., Peircean types VIII, IX, and X) have important visual components in written discourse. All of these sign types, both the verbal and the visual, belong to a single theoretical system built on common principles. These common principles enable a common vocabulary, the classification of visuals in terms of form, reference process, and interpretation type, at first, second, and third levels. Regardless of other vocabulary that visual information designers might choose to use, the consistent Peircean terminology allows a common ground for discussing useful properties of each visual type and how problems arise when any visual type is misapplied in an unsuitable context, exceeding the range of its effective use.

So, for example, the decorative motifs (i.e., "qualisigns") that tend to proliferate in PowerPoint presentations will necessarily interfere with any informational argument because, as shown in Table 1, a presentation that has the form of an argument (3), indexical bullet-point connections to information (2) but only a decorative-level (qualisign) interpretation (1) is ill-formed (*3-2-1) and therefore, as noted above, such presentations are ungrammatical in the system. Likewise, we would classify as ungrammatical a visual with the form *1-2-1, for example, colored tags that are physically adjacent to information or objects as signals but that call undue attention to themselves in their decorative interpretations (recalling Figure 10, chapter 4).

To improve visual communication, then, is to establish the effective function of each visual type relative to communicative contexts, situations in which author and audience needs can be matched with the strengths of particular visual forms and visual media, whether that media consist of printed pages, billboards, PowerPoints, or Web pages, etc. We must further recognize the weaknesses and limitations of any visual type.

What all 10 of these sign categories already have in common with existing visual design discussion is the icon-index-symbol typology, a distinction recognized in most discussions of visual communication, but also the subject of some significant terminological confusion, for example, the proper distinction between an icon and a symbol and the recognition of three distinct subcategories of icon and three distinct subcategories of symbol.

In Peirce's original system, 3 of the 10 sign types are icons, 3 are symbols, and 4 are indices, but these key distinctions are almost invariably overlooked. We have therefore revisited and clarified this three-part taxonomy of icon/index/symbol and have proposed exact criteria for distinguishing the subtypes of each, based on Peirce's own strategy of distinguishing three distinct levels of form, reference process, and interpretation type, each level based on Peirce's primitive categories of Firstness, Secondness, and Thirdness.

UNIVERSAL INFORMATION DESIGN: VISUALS AND TEXT

Peirce wrote, in an unpublished manuscript, that all forms of rhetoric should be understood as the art of making any kind of communication effective.

> Evidently, our conception of rhetoric has got to be generalized; and while we are about it, why not remove the restriction of rhetoric to speech? What is the principle virtue ascribed to algebraic notation, if it be not the rhetorical virtue of perspicuity? Has not many a picture, many a sculpture the very same fault which in a poem we analyze as being "too rhetorical?" (1978, p. 148)

Within the field of technical communication, textual rhetoric is thus the art of making effective texts, within a larger rhetoric that is the general art of making effective visual signs of any type—tables, graphs, diagrams, charts, photographs, and illustrations, but also page layout choices, margins and borders, white space and typefaces. Text is a distinct type of visual within this system, a type that nevertheless usually needs the support and clarification of other types in order to create effective communication.

We can understand effectiveness only in terms of goals: what the author means to accomplish with this or that visual effect and what effect an audience needs when presented with a text supported with visuals. We can therefore begin, at least, to understand author-to-audience goals by again following Peirce's semiotics and sorting these goals into three major heuristic categories: decoratives, indicatives, and informatives. Decorative visuals evoke feelings (which, in terms of phenomenology, is Peircean Firstness); indicative visuals provoke action (Secondness); and informative visuals promote understanding (Thirdness).

What we as authors of this book wanted for ourselves, and what we think will be helpful to professional communicators at large, is language that *explains* why something fails or why a visual is considered effective or not, with, again, effectiveness being achieved (or not achieved) to the extent that textual or other visual-design choices align (or fail to align) with audience needs and authorial goals. The Peircean system of visuals allows us to match form with suitable purpose in a consistent way.

What we have seen instead in the visual design literature (so far) is a best-practices approach to using visuals but not necessarily a unified conceptual framework on which to hang this long list of otherwise rather loosely related assumptions and assertions.

For example, PowerPoint scholars tout using no more than seven bullets per slide and avoiding flashy fly-ins. While these tips are helpful, Peirce's semiotic framework tells us *why* this advice is important: bullets and moving indicators are indexical, they are interpreted by means of physical action. Viewers are physically moving their eyes to follow the bullets, and too many indicatives in a presentation overworks and fatigues the audience. Thus the "best-practice" advice has a principled, systematic explanation. In similar ways, Peirce has provided us with a unified taxonomy for all visuals—a structure on which we can build on and encourage cross-disciplinary exchanges on creating, using, and reading visuals.

Prior to reading this book, you may have held the view that visuals were separate from text in form and in function. Text is to be read; visuals are to be, for the most part, seen. We do not contend here that images, diagrams, tables, and text are the same. We argue, instead, that text and images, diagrams, etc. are *all* examples of visual communication via an application of Peircean semiotics: text as form (that is, typeface choice, etc.) is decorative or iconic, but text as interpretable meaning is informative. Visuals are likewise decorative and/or informative, and clearly, both text and visuals have indicative properties. Peirce postulated "a universal art of rhetoric, which shall be the general secret of rendering signs effect including under the term 'sign' every picture, diagram, natural cry, pointing finger . . . numeral, word, sentence, chapter, book" (1978, pp. 148–149).

What Peirce is talking about here is more than a close integration of the visual and the textual. It is the assertion that rules of visual and textual design all belong to a common system, built on and guided by similar principles. Visual information design is commonly thought to be a fairly new area of critical study, particularly in the field of technical communication, but this view is only an artifact of the historical accidents that have caused Peirce's work to be overlooked for so long.

Why do we need a common language for visuals? For one, it would provide us the means, as a community of professionals working in a common theoretical paradigm, to better assess visuals' effectiveness and to teach how to measure any given visual's effectiveness for our students and colleagues.

INTEGRATING OTHER VISUAL TERMINOLOGIES

To date, several scholars have created their own visual design taxonomies, primarily for the purpose of analyzing and assessing the effectiveness of visuals. For example, Paul Lester's Six Perspectives (2003) encourages readers to examine visuals with a kaleidoscopic six-part perspective: personal, historical, technical, ethical, cultural, and critical. Lester created this seemingly panoramic approach to image analysis to increase visual literacy. In doing this, Lester primarily offers particular audience analyses based on particular graphics. Each analysis in this perspectives-mode offers a unique constellation of personal, historical, and cultural associations, but each offers relatively few useful generalizations applicable to whole classes of visuals.

Expanding on Lester's personal perspective, Rick Williams and Julianne Newton (2007) espouse the Personal Impact Assessment (PIA) approach to assessing visuals, in essence a set of evaluation criteria based on one's personal reaction to a visual. Although we agree that one's own personal reaction to a visual should be taken into account, we would incorporate such response into a larger, more useful visual design system taking other factors into account beyond visual emotability. While Lester as well as Williams and Newton offer insightful approaches, their assessment categories hinge largely on emotional criteria, based on one's personal and social responses to visuals. In other words, their analyses are valuable but touch only the surface as it were, the level-1 layers, of possible visual analysis, as shown in Table 2.

Table 2. Other Visual Terminologies (in Italics) Mapped into the Peircean System[a]

Complexity of **Visual Form**	Connection to **Referent**	Complexity of **Interpretation**
		1. Emotional associations— *e.g., personal, historical, cultural, critical*
	1. Associations similar to prior experience— *e.g., audience, level 1 context*	2. Perceptual or physical actions— *e.g., to notice, to select*
1. Unified gestalts— *e.g., emphasis, organization, color scheme*	2. *Connections directly physical— e.g., alignment, proximity, level-2 context*	3. Propositional assertions true or false, relevant or irrelevant
2. Contrasting phases— *e.g., sequence, juxtaposition*	3. Relations in known-code systems *level-3 context*	
3. Developed Sequential Statements		

[a]There is a general tendency in most visual-design discussion to dwell on emotional associations (1-1-1), imagery (1-1-2), indicative signals (1-2-2), and physical reactions (2-2-2) at the expense of "deeper" aspects of visual communication, i.e., informative, 3rd-level elements.

As noted at the outset, our purpose here is not to supplant any prior approaches to visual communication but only to demonstrate that all of these approaches together can be systematized (and critiqued for gaps in analysis) within the Peircean framework. Alchemists in the Middle Ages made many valid observations about the behavior of individual elements and specific combinations of elements, but until the discovery/invention of the periodic table, there was no effective unification, systematization, and explanation of those observations. There was no science of chemistry.

Offering a more goal-driven approach to understanding visuals, Ball and Arola's (2005) *IX techcomm: Visual Exercises for Technical Communication* provides definitions of visual concepts and assignments on how to analyze based on those concepts, which include text and purpose, element and context, audience, contrast and color, emphasis and framing, alignment, proximity, organization, and sequence. Although these rubric criteria are more specific and concrete than a personal response approach, the visual reader is nevertheless relying on an unorganized list of possible considerations that one might use to assess any visual's effectiveness. The list is "flat" in that it fails to separate global issues such as purpose from relatively local and specific elements like color, alignment, or sequence.

Alchemists made a similar error in confusing general states of matter (solid/earth, liquid/water, gas/air, plasma/fire) with specific substances (silicon, H_2O, oxygen, etc.)

Not surprisingly, we've found that most visual taxonomies have been based on the rhetorical tradition, little changed since Aristotle (Handa, 2004; Hill & Helmers, 2004; Kostelnick & Hassett, 2003). In this same vein, Kostelnick and Roberts's *Designing Visual Language* (1997) centers on traditional Aristotelian categories of arrangement (emphasis, clarity, conciseness, tone, and ethos). Again, while these approaches have merit, none of them offer a comprehensive, unified, and potentially universal framework for all visual types, including text.

In terms of visuals, hand-painted, primarily decorative illustrations were the norm when classical rhetoric was developed, and these illustrations occupied a middle ground between physical and conceptual resemblance in an era that predated both photographs and extensive technical diagramming. We argue, therefore, for an underlying logic of visual representation, one that needs to be more current and cognizant of modern visual and textual forms, and one that would be at least as helpful as classical rhetoric to explain why an effective diagram (Peirce type 1-1-3) has properties of emphasis, clarity, tone, etc. that are distinctly different from properties of effective images (1-1-2; see Figure 2).

Today, the image/diagram distinction is utterly critical. For example, a high-detail resolution in a photograph is usually helpful, but increased conceptual detail in a diagram tends to be distracting, creating problems of relevance, relevance being a level-3 issue governing the interpretation of diagrams, as opposed to level-2 issues governing the interpretation of images.

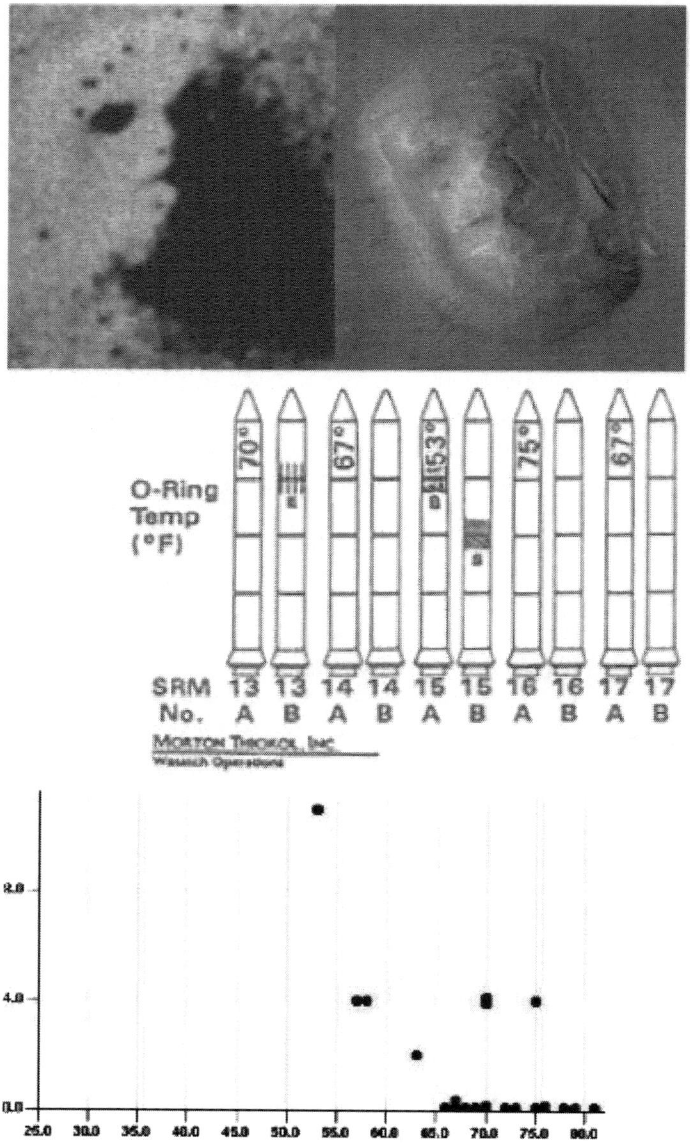

Figure 2. Data/ink ratio rule (Tufte) not applicable to images.

Images' indicative function is improved with extra detail. Diagrams' informative function is improved with less detail, selected for relevance. The data/link ratio rule is specifically a rule for informative visuals, not decoratives or indicatives.

In Figure 2 (top), for example, the question of whether the "Face" on Mars was an alien, megalithic sculpture or a natural rock formation could be put to rest only by a higher-resolution photograph than was originally available (nasa.gov). In contrast, a diagram is generally more useful if it contains only the information relevant to a specific proposition (as Edward Tufte has so eloquently argued), applying hindsight to data available before the Challenger disaster (1997, p. 45–48; see Figure 2, bottom). What is missing, however, even in Tufte's thinking, is a recognition that effective images serve a different configuration of communicative purposes and constraints than diagrams do. Figure 2 demonstrates how effective images (in their indicative function) are not subject to the same data/ink ratio rules that Tufte convincingly applies to diagrams (2001).

In terms of Table 2, we find the classical-rhetoric approaches penetrating through level-1, felt-perception layers to the level-2, indicative-action layers, but generally falling short, still, of the level-3 informative layers. This insufficiency in classical rhetoric also makes sense when we realize that the classical rhetoricians of ancient Greece and Rome were primarily focused on moving audiences to political action, whereas the pre-scientific information culture was in its infancy at best. So, when we consider visual analyses in the classical-rhetoric mode, we typically find this gap in the discussion, a lack of precision in differentiating informative-visual elements.

To take another example, Ball and Arola discuss "symbols" (more precisely, icons with some symbolic elements) associated with clothing care (Figure 3). These are presented as examples of elements and context. "Context is additional information about a text, such as where it is located, how it is meant to be read, or what surrounds it. . . . Context is a broad concept—it also includes the cultural knowledge that an audience brings to a text" (2005, sect. 2 on CD).

The difficulty here is that the classical heading of context conflates very different things: the physical location of the icons/symbols, their intended purpose, and "cultural knowledge" about them. "Cultural knowledge" here also conflates two things that ought to be kept separate, at least in the Peircean analysis; that is, the iconic case where readers are acculturated to recognize similarities they otherwise wouldn't notice, and on the other hand, the genuine symbolic case where readers are acculturated not to expect form-object similarities at all but rather systematic contrasts between symbolic elements in a coded paradigm.

In the clothing-care examples, the triangle glyph meaning "bleach" is a genuine symbol, as is the circle meaning "dry cleaning." Most other glyphs in the system (cf. http://www.sdahq.org) are still recognizably iconic, even if the form-object similarity has to be learned. Our major point here is that the Peircean system enables us to separate and precisely delineate what is otherwise quite vague in classical rhetoric: three levels of context as shown in Table 2 (center column). Purpose is meanwhile a more global consideration, but the most effective purpose

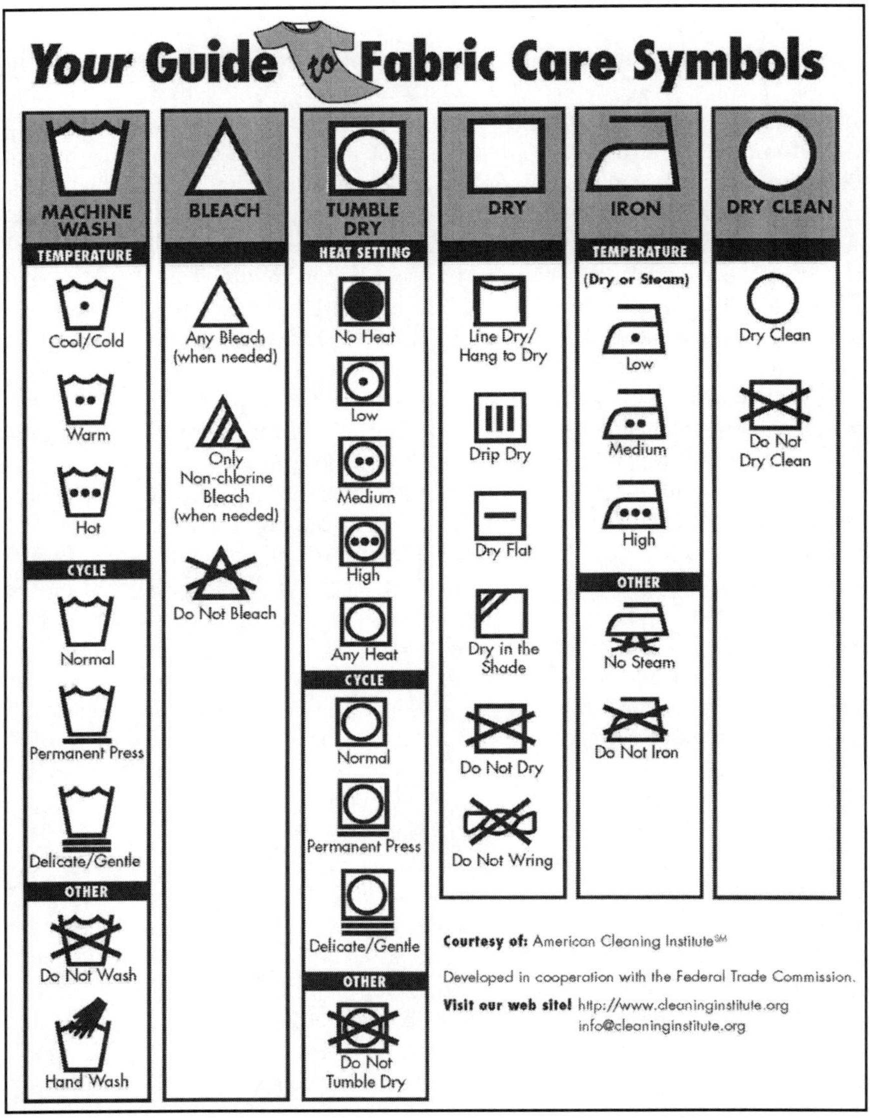

Figure 3. Icons and symbols combined in one visual system.

Iconic interpretation by similarity and symbolic interpretation by code require different kinds of "cultural context." Peircean analysis allows us to keep those diverse types of "context" distinct, even as they are mixed together in one visual.

of any given visual also can be calculated with some precision in the Peircean system as laid out in Table 2.

For example, the clothing-care glyphs (level 1 in column 1) are on tags physically attached to clothing (level 2 in column 2). The glyphs do not represent the clothing but rather what actions are to be taken (or avoided) with the clothing (level 2 in column 3). These facts allow us to precisely classify the glyphs, overall, as 1-2-2 visuals. They are, in other words, neither icons nor symbols in terms of their purpose, but rather indicative signals with icon-like form (recalling the arrow-shaped bullets in many PowerPoint presentations).

CONCLUSION

Our overall philosophy is perhaps most similar to Kress and van Leeuwen's (1996) idea of the social semiotics of truth in their book *Reading Images: The Grammar of Visual Design*. We share their zeal for understanding how meaning is made and classified in images through semiotics, but we would further argue that a mature science of effective graphics and ethical graphical choices is needed, beyond the impressionistic, associative, and emotional reading of images, just one visual type out of many. Any mature science of the visuals needs to, in one way or another, provide a taxonomy of different visual types and specifically be able to weigh each type against author purpose, audience expectation, media choice, and shared author-audience goals.

In sum, we are likewise interested in visual language, as are these scholars cited above, but more particularly, we wish to connect visuals' core formal properties with their core purposes, to motivate and clarify strategies for constructing more effective and efficient visual communication. To do so, we turn to a semiotic analysis, aligned with what is essentially a grammatical analysis (in the linguistic sense, where grammatical visuals = visuals well-formed according to form/content rules). In doing this, we aspire to provide, with a renewed discussion and analysis of Peirce's seminal work, a more comprehensive understanding of the full range of semiotic language that governs the effective functioning of visuals.

References

Abes, C. (2001). Bring data to life: Art and information can complement each other. *Macworld.com* [sumo graphic]. Retrieved August 5, 2008, from http://www.macworld.com/article/2270/2001/06/howtodata.html

Adams, S. (2011). *Dilbert strip dialogue.* Retrieved February 6, 2011, from http://www.dilbert.com/strips/comic/2011-02-06/

Alley, M. (2003). *The craft of scientific presentations: Critical steps to succeed and critical errors to avoid.* New York: Springer.

Alley, M. (2008). *Writing guidelines for engineering and science students.* Retrieved from http://writing.engr.psu.edu/

Alley, M., & Neeley, K. (2005). Rethinking the design of presentation slides: A case for sentence headlines and visual evidence. *Technical Communication, 52*(4), 417–426.

Amare, N., & Manning A. (2006). *A usability model of hypertext based on the semiotics of C. S. Peirce.* In proceedings of the International Professional Communication Conference, pp. 47–56.

Amare, N., & Manning, A. (2007). The language of visuals: Text + graphics = visual rhetoric. *IEEE Transactions on Professional Communication, 50*(1), 57–70.

Amare, N., & Manning, A. (2008). *A language for visuals: Design, purpose, usability.* In proceedings of the International Professional Communication Conference [CD ROM].

Amare, N., & Manning, A. (2009a). *Emotion-spectrum response to form and color: Implications for usability.* In proceedings of the International Professional Communication Conference. Article #42. Piscataway, NJ: IEEE.

Amare, N., & Manning, A. (2009b). Editor-author ethics: Examining real-world scenarios from interviews with three journal editors. *Journal of Technical Writing and Communication, 39*, 285–303.

Amare, N., Nowlin, B., & Weber, J. (2011). *Technical editing in the 21st century.* Boston: Prentice Hall.

Ball, C., & Arola, K. (2005). *IX tech comm: Visual exercises for technical communicators.* [CD ROM]. New York: Bedford-St. Martins.

Barton, B., & Barton, M. S. (2004). Ideology and the map: Toward a postmodern visual design practice. In J. Johnson-Eilola & S. Selber (Eds.), *Central works in technical communication* (pp. 232–252). Oxford, UK: Oxford University Press.

Bernhardt, S. (1986). Seeing the text. *College Composition and Communication, 37,* 66–78.

Blokzijl, W., & Andeweg, B. (2007). *The effect of text slides compared to visualizations on learning and appreciation in lectures.* In proceedings of IPCC Conference [CD ROM] 10.1109/IPCC.2007.4464074.

Boekelder, A., & Stëehouder, M. (1998). Selecting and switching: Some advantages of diagrams over tables and lists for presenting instructions. *IEEE Transactions on Professional Communication, 41*(4), 229–241.

Brent, J. (1998). *C. S. Peirce: A Life*. Bloomington: Indiana University Press.

Brumberger, E. (2003). The rhetoric of typography: The persona of typeface and text. *Technical Communication, 50,* 206–223.

Burch, R. (2008). Charles Sanders Peirce. In E. N. Zalta (Ed.), *The Stanford encyclopedia of philosophy*. Retrieved September 10, 2008, from http://plato.stanford.edu/archives/fall2008/entries/peirce/

Burks, A. W. (Ed.). (1958). *Collected papers of C. S. Peirce* (Vols. 7–8). Cambridge, MA: Harvard University Press.

Campbell, K. (1995). *Coherence, continuity, and cohesion: Theoretical foundations for document design.* Mahwah, NJ: Lawrence Erlbaum.

Clark, D. (2008). Content management and the separation of presentation and content. *Technical Communication Quarterly, 17*(1), 35–60.

Collier, G. (1996). Affective synesthesia: Extracting emotion space from simple perceptual stimuli. *Motivation and Emotion, 20*, 1–32.

Collier, G., Dorflinger, G. K., Gulick, T. A., Johnson, D. L., McCorkle, C., Meyer, M. A. et al. (1976). Further evidence for universal color categories. *Language, 52*(4), 884–890.

Cowan, N. (2001). The magical number 4 in short-term memory: A reconsideration of mental storage capacity. *Behavioral and Brain Sciences, 24*(1), 87–114.

Doumont, J. L. (2002). Magical numbers: The seven-plus-or-minus-two myth. *IEEE Transactions on Professional Communication, 45*(2), 123–127.

Doumont, J. L. (2005). The cognitive style of PowerPoint: Slides are not all evil. *Technical Communication, 52*(1), 64–70.

Dragga, S. (1996). Is this ethical?: A survey of opinion on principles and practices of document design. *Technical Communication, 43*, 255–265.

Dragga, S., & Voss, D. (2001). Cruel pies: The inhumanity of technical illustrations. *Technical Communication, 48*, 265–274.

Dragga, S., & Voss, D. (2003). Hiding humanity: Verbal and visual ethics in accident reports. *Technical Communication, 50*, 61–82.

Elliot, D., & Lester, P. (2002). Visual communication and an ethic for images. In K. Smith et al. (Eds.), *Visual communication research.* Westport, CT: Praeger. Retrieved from http://commfaculty.fullerton.edu/lester/writings/imageethic.html

Ellis, H. C., Ottaway, S. A., Varner, L. J., Becker, A. S., & Moore, B. A. (1997). Emotion, motivation, and text comprehension: The detection of contradictions in passages. *Journal of Experimental Psychology, 126*(2), 131–146.

Evans, P., & Thomas, M. (2008). *Exploring the elements of design* (2nd ed.). Clifton Park, NY: Delmar Cengage Learning.

Ewalt, D. M. (2005, October 6). Interview with Stan Lee. *Forbes.com.* Retrieved November 3, 2005, from http://www.forbes.com/2005/10/19/lee-stan-comic-superhero-comm05-cx_de_1024lee.html

Forster, E. M. (1941). *Aspects of the novel*. London: Edward Arnold.

Fulkerson, R. (1979). Four philosophies of composition. *College Composition and Communication, 30,* 343–348.

Gaskins, R. W. (1996). That's just how it was: The effect of issue-related emotional involvement on reading comprehension. *Reading Research Quarterly, 31*(4), 386–405.

Handa, C. (Ed.). (2004). *Visual rhetoric in a digital world: A critical sourcebook.* New York: Bedford-St. Martins.

Hartshorne, C., & Weiss, P. (Eds.). (1935). *Collected papers of C. S. Peirce* (Vols. 1–6). Cambridge, MA: Harvard University Press.

Helmers, M. (2006). *The elements of visual analysis.* New York: Pearson Longman.

Hill, C., & Helmers, M. (Eds.). (2004). *Defining visual rhetorics*. Philadelphia: Lawrence Erlbaum.

Iversen, M. (1986). Saussure v. Peirce: Models for a semiotics of visual art. In A. Rees & F. Borzello (Eds.), *The new art history* (pp. 82–94). London: Camden Press.

Jennings, A. (2007). *The sales aspect of engineering presentations.* In proceedings of the International Professional Communication Conference, 10.1109/IPCC.2007.4464082.

Kimball, M., & Hawkins, A. (2008). *Document design: A guide for technical communicators.* Boston/New York: Bedford/St. Martin's.

Kostelnick, C., & Hassett, M. (2003). *Shaping information: The rhetoric of visual conventions*. Carbondale: Southern Illinois University Press.

Kostelnick, C., & Roberts, D. (1997). *Designing visual language: Strategies for professional communicators.* White Plains, NY: Longman.

Kress, G., & van Leeuwen. T. (1996). *Reading images: The grammar of visual design.* New York: Routledge.

Lester, P. (2003). *Visual communication: Images with messages* (3rd ed.). Australia: Wadsworth.

Lester, P. (2004). Living with pornography: An essay of exactly 1,000 words. *journalism.org.* Retrieved from http://commfaculty.fullerton.edu/lester/writings/1000pictures.html

Mackiewicz, J. (2006). *Audience perceptions of typefaces in projected PowerPoint text slides*. In proceedings of the National Professional Communication Conference, 2006 IEEE, pp. 68–76.

Mackiewicz, J., & Moeller, R. (2004). *Why people perceive typefaces to have different personalities.* In proceedings of the International Professional Communication Conference, pp. 304–313.

Manning, A. (1989). The invariant code-significance of lexical items. *Semiotica, 73,* 101–119.

Manning, A., & Amare, N. (2005). *Using visual rhetoric to avoid PowerPoint pitfalls.* In proceedings of the International Professional Communication Conference, pp. 281–287.

Manning A., & Amare, N. (2006). Visual-rhetoric ethics: Beyond accuracy and injury. *Technical Communication, 53*(2), 195–211.

Marty, R. (1990). *L'algebra des signes*. Amsterdam: John Benjamins.

McCloud. S. (1994). *Understanding comics: The invisible art.* New York: Harper Paperbacks.

McKee, R., & Bronwyn, F. (2003, June 1). Storytelling that moves people. *Harvard Business Review.*

Microsoft, Inc. (2003). *Have a great idea? Make a persuasive pitch*. Retrieved June 25, 2008, from office.microsoft.com/en-us/help/HA010838371033.aspx

Miller, G. (1956). The magical number seven, plus or minus two: Some limits on our capacity for processing information. *The Psychological Review, 63*, 1–97. Retrieved March 20, 2008, from www.musanim.com/miller1956

Moriarty, S. (2005). Visual semiotics theory. In K. Smith, S. Moriarty, K. Kenney, & G. Barbatsis (Eds.), *Handbook of visual communication: Theory, methods, and media*. Philadelphia: Lawrence Erlbaum.

Newton, J. (2005). Visual ethics theory. In K. Smith, S. Moriarty, K. Kenney, & G. Barbatsis (Eds.), *Handbook of visual communication: Theory, methods, and media*. Philadelphia: Lawrence Erlbaum.

Ou, L., Luo, M. R., Woodcock, A., & Wright, A. (2004). A study of colour emotions and colour preference. Part I: Color emotions for single colours. *COLOR Research and Application, 29,* 232–240.

Peirce, C. S. (1978). Ideas, stray or stolen about scientific writing, no. 1. *Philosophy and Rhetoric, 11,* 147–155.

Ransdell, J. (1998). Who is Charles Sanders Peirce? *Peirce Telecommunity Project*. Retrieved October 4, 2008, from www.cspeirce.com/whoiscsp/whoiscsp.htm

Reaves, S., Hitchon, J. B., Park, S., & Yun, G. (2004). If looks could kill: Digital manipulation of fashion models. *Journal of Mass Media Ethics, 19*(1), 56–71.

Riley, K., & Parker, F. (1998). Parallels between visual and textual processing. *IEEE Transactions in Professional Communication, 41*(3), 175–185.

Robertson, J. (1994). A Peircean categorical analysis of the English morphemes -ing, -ed, and -s. *Semiotica, 102,* 179–223.

Scholes, R. (1983). *Semiotics and interpretation*. New Haven, CT: Yale University Press.

Schriver, K. (1997). *Dynamics in document design*. New York: Wiley.

Sedlack, R., Shwom, B., & Keller, K. (2006). *Graphics and visual communication for managers*. Mason, OH: Thompson/South-Western.

Shaw, G., Brown, R., & Bromiley, P. (1998, May 1). Strategic stories: How 3M is rewriting business planning. *Harvard Business Review*.

Sternberg, R. (2003). *Cognitive psychology* (3rd ed.). Australia: Thomson Wadsworth.

Tufte, E. (1997). *Visual explanations: Images and quantities, narrative and explanation*. Cheshire, CT: Graphics Press.

Tufte, E. (2001). *The visual display of quantitative information* (2nd ed.). Cheshire, CT: Graphics Press.

Tufte, E. (2003). *The cognitive style of PowerPoint*. Cheshire, CT: Graphics Press.

van Hooijdonk, C., de Vos, J., Krahmer, E., Maes, A., Theune, M., & Bosma, W. (2007). *On the role of visuals in multimodal answers to medical questions*. In proceedings of the International Professional Communication Conference 10.1109/IPCC.2007.4464073.

Walsh, T. (1998). Illustrations in oral presentations: Photographs. *IEEE Transactions on Professional Communication, 41,* 209–212.

Williams, R. (2008). *The non-designer's design book* (3rd ed.). Berkeley, CA: Peachpit Press.

Williams, R., & Newton, J. (2007). *Visual communication: Integrating media, art, and science*. Mahwah, NJ: Lawrence Erlbaum.

Wright, R. (2002, January 7). Vintage auto ads reflect some surprising vintage attitudes. *Joyrides*. Retrieved August 5, 2008, from http://info.detnews.com/joyrides/story/index.cfm?id=295

Xin, J., Cheng, K. M., Taylor, G., Sato, T., & Hansuebsai, A. (2004). Cross-regional comparison of colour emotions. Part II: Qualitative analysis. *COLOR Research and Application, 29,* 458–466.

Zentner, M. (2001). Preferences for colours and colour-emotion combinations in early childhood. *Developmental Science, 4,* 389–398.

Index

Action triggers, 115–117, 119, 124
Action triggers, shorthand code, 126*t*
  elevated, 145–146, 148
  interpretation type, 126–128
  misuse, overuse, 129
  require contrast, 131
  shorthand code, 126*t*
  web page links, 129
Advertisements, cultural context, associative connections, 40*f*
Alley, M., 64
Alternative unification models, 25
Animation, indicative (watching) *vs.* ritual-sequence (learning), 144*f*
Architecture example images, 108*f*
Aristotelian categories, 195
Arola, K., 195

Ball, C., 195
Borders, backgrounds, images, 8
Bullet marks, 9
Bulleted lists, 8–10
Bullets and links, unorganized, 116*f*

Center for Disease Control (visual), 5–6
Chartjunk examples, 100*f*
Coded-system relationships, 169, 175–176, 177*f*
Color, feeling, generic associations, 54*t*
Communication development, 77, 103, 128
Community of inquiry, 16
Composed form-feeling associations, 50*f*
Compromised visuals, illustrative examples, 4–7
Containment, 24
The Craft of Scientific Presentations (Alley), 64

Data, without image distraction, 73*f*
Data/ink ratio rule, 196*f*
Decorated graphic, 29*f*
Decorated persuasive graphic, 34*f*
Decorative, indicative, informative, visual problems, revisions, 5*f*
Decorative design parameters, Peircean, 35–36
  advertisements, cultural context, 40f
  basic form, 37
  clashing *vs.* unified decoratives, 38–39f
  first level, 41
  informative reconstruction, 42*f*
  interpretation type, 37
  key design parameters, 43*t*
  reference process, 37
  10-class system of visual types, 43*f*
Decorative effects in images, 80
  form/feeling associations, 81*t*
  variety, contrast, color effects, 82, 83*f*

Decorative/indicative forms elevated to informative forms, 146*f*
Decorative-indicative images, 18–19
Decorative-informative graphics, 19
Decorative matching problems, 47*f*
Decorative purpose, applications, limitations
  color, typefaces, 51–54
  composed form-feeling associations, 50*f*
  ethical action, 57–58
  feeling, 47–48, 52
  generic associations, color, feeling, 54*t*
  generic associations, typeface, feeling, 54*t*
  generic associations of form, feeling, 48*t*
  interrupted, inconsistent schemes, 52*f*
  matching problems, 47*f*
  movie poster color-form compositions, 56*f*
  primary form-feeling associations, 49*f*
  shape, texture, 54–57
  similarity, 46, 52
  toward a decorative ethic, 57–59
  typeface, 57
  typeface, color inconsistent with message, 53*f*
  unity, 44, 51
Decorative(s), 8
  definition, 28
  integrated with indicatives, 134*f*
  key design parameters, 43*t*
  potentially misleading, 36*f*
  primary function, 28
  reference process, 37
  ugly, emotional-clash graphic, 35*f*
  used problematically as indicatives, 132*f*
Designing Visual Language (Kostelnick, Roberts), 195
Diagram, applications, limitations
  architecture example images, 108*f*
  detail management, 110–112
  essential distinctions between images, diagrams, 114*t*
  general ideas, 112
  intermediate visualization, 106
  interrupted patterns, 45*f*
[Diagram, applications, limitations]
  mediating function of informative visuals, 105*f*
  mediation between language, perception, 105–110
  overly detailed *vs.* stripped-down diagram., 111*f*
  photograph *vs.* diagram, 113*f*
  single correct interpretation for diagrams, 109*f*
Diagram design parameters, Peircean
  basic form, 97
  chartjunk examples, 100*f*
  diagrams relative to other visual types, 104*f*
  information graphics, 101
  interpretation type, 101–104
  key design parameters of diagrams, 103*t*
  photos of materials with diagrammatic function, 98, 99*f*
  referential process, 97–101
  well-formed, difficult-to-interpret diagrams, 102*f*
Diagrams, 10, 93–95
  effective, 110
  image *vs.* diagram, 96*f*
  information-perception bridges, 104
  overly detailed *vs.* stripped-down, 111*f*
  photograph *vs.*, 113*f*
  relative to other visual types, 104*f*
  single correct interpretation, 109*f*
  well-formed, difficult-to-interpret, 102*f*
Disciplined thought, 15–16
Disordered textual decoratives, 168*f*
Dynamic action triggers *vs.* static signals, 125*f*
Dynamic cascade of visual types, 130*f*

Earthrise photo, multiple captions, meanings, 66*f*
Effectiveness, 11, 192
Emotional effect, 32
Emotional message, 30
Essential distinctions between images, diagrams, 114*t*

Ethic(s)
decorative, 57–59
diagram, 112–114
epistemological, 11–17
ethical abuse, 84
ethical action, 15–16, 57–58, 88–89
image, 88–91
index, 162–163
indicative, 139–140
issues, 4–5
unethical visual, 2
visual language, 185

Feeling creation, 32
Firstness
defined, 21–23, 26
forms, colors, 33
Food and Drug Administration (FDA), 5–6, 120*f*
fda.gov, fewer bullets, more effective, 120*f*
Form, feeling, generic associations, 48*t*

Generic associations,
form, feeling, 48*t*
typeface style, feeling, 54*t*
Gestalt principles, 44–45
Goal, 8
Grammar, 20
Peircean categories, 21–24, 64
traditional, 173
visual, 103, 131
Graphic
undecorated, 31*f*
undecorated persuasive, 33*f*

Hedonist agenda, 12–13
Horizontal/vertical sequencing, web page tabular structure, 150*f*

Icons
range of, 87*f*
symbols combined, 198*f*

If-then statement, 24
Illustrative examples, image-to-diagram, signal-to-index, action-to-ritual, 147*f*
Image applications, limitations, 79
action mediating effects, 84
decorative effects, 80–83
image-driven message, 85*f*
Image design parameters, Peircean
basic form, 72–74
interpretation type, 74–78
referential process, 74
Images
adjacent information and, 135*f*
data without image distraction, 73*f*
deception, 70
decoratives *vs.* images *vs.* diagrams, 63*f*
definition, 61
diagram *vs.*, *96f, 195*
distorting, substituting, distracting from actual information, 71*f*
earthrise photo, multiple captions, meanings, 66*f*
feeling creation, 65
icons, 62
image-based deception, 71
image-driven message, 85*f*
image-driven political message, 86*f*
image ethic, 88–91
image reference, similarity to prior experience, 76*f*
informative graphics, image-cluttered, 69*f*
informative graphics, image-removed, 70*f*
key design parameters, 77*t*
legitimate use, 72
limitations, alternatives, 85–88
overuse, unethical use, 72
public service message with image, 67*f*
public service message without image, 68*f*
range of icon types, 87*f*
similarity to prior experience, 76*f*
variety, contrast, color effects, 82, 83*f*

Indicative design parameters, Peircean
action triggers, 126*t*
basic form, 121–124
bullet point, animation difficulties, 136–139
decoratives used problematically as indicatives, 132*f*
dynamic cascade of visual types, 130*f*
images that fail to match adjacent information, 135*f*
indicative elements, organized, 123*f*
indicative elements, unorganized, 122*f*
indicatives relative to other visual types, 127*f*
interpretation type, 126–129
referential process, 124–126
signal indicatives, 126*t*
static signals *vs.* dynamic action triggers, 125*f*
Indicative photograph, manipulated for decorative effect, 80*f*
Indicative(s), 8–9
action trigger(s), 115–117, 121
applications, limitations, 129–139
bullet points, 115, 124, 140
containing information content, 19–20
design choices, 124
elements, organized, 123*f*
elements, unorganized, 122*f*
fda.gov, fewer bullets, more effective, 120*f*
four kinds, 118–119
meaning, 118
organized, 123*f*
overuse, 117, 129
pointing to information content, 19
pure signals and actions, 117*f*
reference indices, 117, 119
relative to other visual types, 127*f*
ritual sequences, 117, 119
signal(s), 115–118, 129, 165
unorganized, 122*f*
unorganized lists of links, 116*f*
web page links, 22, 119, 130, 138–139
Indices
informative, 141–143
reference, 143, 145, 151
[Indices]
relative to other visual types, 156*f*
ritual-sequence, 143, 145, 151
Information design
alternative unified theories, 25
Peircean categories, 3
practical implications, 155
problems, revisions, 6
Informative graphics
image-cluttered, 69*f*
image-removed, 70*f*
Informative index design parameters, Peircean
basic form, 151–152
ethical improvements, 162–163
indexical applications, limitations, 157–162
indices relative to other visual types, 156*f*
interpretation type, 153–157
postage calculation, familiar routing/ritual, 162*f*
reference indices, key design parameters, 153*t*
referential process of informative indices, 154*f*
ritual-sequence indices, key design parameters, 153*t*
Informative indices, 141, 143, 145
animation, indicative (watching) *vs.* ritual-sequence (learning), 144*f*
decorative/indicative forms elevated to informative forms, 146*f*
horizontal/vertical sequencing, web page tabular structure, 150*f*
illustrative examples, image-to-diagram, signal-to-index, action-to-ritual, 147*f*
informative indices pointing to larger informative structure, 148*f*
referential process, 154*f*
socially coded, 142
unordered indicatives *vs.* informative reference index, 142*f*
Informative language, 20
Informative reconstruction of overly decorated visual, 42*f*

Informative reference index *vs.* unordered indicatives, 142*f*
Informatives, 8–9
Interrupted, inconsistent schemes, 52*f*
Interrupted patterns, 45*f*

Jennings, A., 64

Key design parameters of decoratives, 43*t*
Key design parameters of images, 77*t*
Kostelnick, C., 195
Kress, G., 199

Language design parameters, Peircean
- coded-system relationships, 169, 177f
- informative relationships, 169
- interpretation, 176
- interpretation type, 176–178
- proposition development, 169, 171, 174
- reference process, 175–176
- text intelligibility optimization, 170, 171*f*

Language forms relative to other visual types, 179*f*
Logical diagram, 94

Magical Number Seven Plus or Minus Two (Miller), 119
Marketing techniques, 12
McCloud, Scott, 86
Measured, 119, 120
Mediating function of informative visuals, 105*f*
Memory, channel capacity, 119–120
Miller, G., 119
Movie poster color-form compositions, 56*f*
Multiple captions, meanings, 66*f*

Navigational menu
- image-cluttered, 90*f*
- image-removed, 91*f*

Newton, Issac, 25

Pathos, 32
Paul Lester's Six Perspectives, 193
Peircean analysis, 6–7, 12–13, 19–20
Peircean model, 6, 165
Peircean parameters
- decorative design, 35–43
- diagram design, 97–104
- image design, 72–78
- indicative design, 121–129
- informative index design, 151–162
- language design, 169–178

Peircean semiotics, 193
Peircean system, book chapter progression, 18*f*
Peirce's 10-class system of visual types, 43*f*
Peirce's categories
- firstness, secondness, thirdness, 24
- grammar, 21–24, 22*f*, 64
- simplication, 190*f*
- visual design, 8–11

Peirce's system, translated, 188
Perception/action face of the Peircean pyramid, 180*f*
Periodic table, 4, 21, 195
Personal Impact Assessment (PIA), 193–194
Photograph *vs.* diagram, 113*f*
Photos of materials with diagrammatic function, 98, 99*f*
Phrasal parallelism, visual unity strategy, 181*f*
Physical adjacency, 127*f*
Physical-object connection, 74
PIA. *See* Personal Impact Assessment
Pie charts, 34
Plain Language movement (U.S. government), 5–6
Postage calculation, familiar routing/ritual, 162*f*
Precedence, 24
Primary form-feeling associations, 49*f*
Primary visual-communication goals, 9*f*
Proposition development, 169, 183–185
- argument, 174
- text-form rule, 173

Public service message
with image, 67*f*
without image, 68*f*
Pure decoratives *vs.* images *vs.* diagrams, 95*f*
Pure signals and actions, 117*f*
Purely decorative (aesthetic elements), 18–19

Range of icon types, 87*f*
Reading images: The Grammar of Visual Design (Kress, van Leeuwen), 199
Reference by physical adjacency, 127*f*
Reference indices, key design parameters, 153*t*
Referential process of informative indices, 154*f*
Relative clarity of complete sentence captions, 184*f*
Removal test, 30
Rhetoric, 11, 192
Ritual-sequence indices, key design parameters, 153*t*
Ritual sequences, 114, 117, 142–143
Roberts, D., 195

Sales Aspect of Engineering Presentations, The (Jennings), 64
Secondness, defined, 21–23, 26
Sentence captions, 184*f*
Sentence intelligibility, overly long subjects, 182*f*
Shorthand code, 41, 74, 77
Sign
defined, 25
types, 3–4, 117, 187–188
Signals, 115–117, 119, 121
indicative, 124, 129, 140
ritual, 4
signal indicatives, shorthand code, 126*t*
social, 105
visual, 1, 139, 176
Similarity, qualities of, 37, 74, 97
Single correct interpretation for diagrams, 109*f*
Static signals *vs.* dynamic action triggers, 125*f*
System of visual types, shorthand code, 78*f*

Text. *See also* Words, sentences, text
essential visual indicatives removed, 166*f*
optimized intelligibility, 170, 171*f*
as visual form, 1
Thirdness, defined, 22–23, 26
Three-dimensional typology (Peirce), 8–11
Tufte, Edward, 97
Typeface, 57
choices, 8, 10
color inconsistent with message, 53*f*
generic associations, styles, feeling, 54*t*

Undecorated graphic, 31*f*
Undecorated persuasive graphic, 33*f*
Understanding Comics (McCloud, Scott), 86
Unethical visual, 2
Unity, qualities of, 72
Universal terminology, visual types grammar, 20
Unordered indicatives *vs.* informative reference index, 142*f*
Utilitarianism, 12–14

Van Leeuwen, T., 199
Vandalized images, disputed unity, 75*f*
Viewer fatigue, 138, 139
Visual communication goals, 8–10
Peirce's primary categories, 9*f*
Visual design
current pedagogy, 2
purpose, vocabulary of, 27
strategies, 27

[Visual design]
traditional, 115
unified theoretical approach, working definitions, 2
Visual design problems
decorative, 5–6
ethical issues, 4–6
indicative, 5
Visual Display of Quantitative Information, The (Tufte, Edward), 97
Visual Exercises for Technical Communication (Ball, Arola), 195
Visual language, implications, applications, limitations
parallelism, 179
perception-action cascade, 178
perception/action face of the Peircean pyramid, 180*f*
phrasal parallelism, visual unity strategy, 181*f*
proposition development, 183
relative clarity of complete sentence captions, 184*f*
sentence intelligibility hindered by overly long subjects, 182*f*
[Visual language, implications, applications, limitations]
Visual types system, shorthand code, 78*f*
Visual(s)
effective, ethical, 2
ineffective, 2
language, 8, 165, 178
purposes, 27
signals, 1, 139, 176
terminologies, other, 194*t*
types, 4
unethical, 2
Vocabulary of purpose, 27

Well-formed, difficult-to-interpret diagrams, 102*f*
Words, sentences, text, 165–186
disordered textual decoratives, 168*f*
text with essential visual indicatives removed, 166*f*
visual language, implications, applications, limitations, 178–185

About the Authors

Nicole Amare is associate professor of technical communication at the University of South Alabama, where she teaches composition, technical writing, editing, ethics, stylistics, and grammar. She has written *Real Life University*, a college success guide; and has edited *Global Student Entrepreneurs*, *Beyond the Lemonade Stand*, and *Giving Back*. Some of her research has appeared in the *Journal of Technical Writing and Communication, Business Communication Quarterly*, *IEEE Transactions on Professional Communication, Women & Language*, and *Technical Communication*. Her most recent book, with Barry Nowlin and Jean Hollis Weber, is *Technical Editing in the 21st Century* (2011, Prentice Hall). She can be reached at namare@usouthal.edu

Alan Manning is a professor of linguistics and English language at Brigham Young University. He teaches graduate courses in writing and research design, and undergraduate courses in linguistics and editing. He is a co-author of *Revising Professional Writing in Science and Technology, Business, and the Social Sciences* (2nd ed., 2007, with Riley, Campbell, and Parker, Parlay Press). He can be reached at alan_manning@byu.edu